The
Practical
Queen Breeder

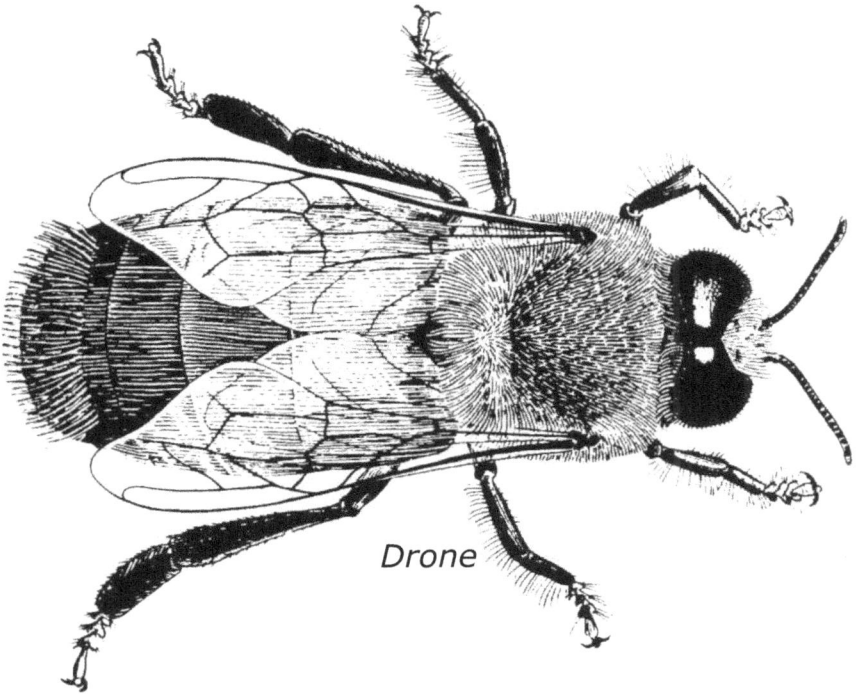

Drone

By Michael Bush

The Practical Queen Breeder—Beekeeping Naturally

Cover Photo of drone © 2011 Alex Wild
www.alexanderwild.com

ISBN 978-161476-076-4

X-Star Publishing Company

Nehawka, Nebraska, USA
http://xstarpublishing.com

210 pages

114 illustrations

Acknowledgments

I would like to thank Kirk Webster for reinforcing and reassuring me on some of the ideas I already had on breeding. Though his methods are quite different he so eloquently has pointed out the pitfalls.

Foreword

I suppose everyone is wondering why the cover and the inside cover have pictures of a drone instead of a queen. I think the importance of the drones is often overlooked so I thought I would draw attention from the beginning.

I also want to point out that at the end of chapters on otherwise blank pages I often put a picture of something I think is of interest but not necessary related to the chapter. I hope this isn't confusing and I hope it presents things that are of general interest.

I hope you will use this book successfully. There are many ways to raise queens. I've tried most of them. The methods laid out here are the ones I'm currently using. I do keep changing them and adjusting but also improving. I notice that between Queen Rearing Simplified and Better Queens, Jay Smith made many adjustments, yet both books are very useful and insightful. I hope this book continues to be useful and insightful into the future even though I will probably make more adjustments as the years go by.

Lighting Burlap

Table of Contents

About this book

My intent for this book is a step by step of how I do queen breeding and queen rearing. Some of this will be *why* I do it, but mostly it will be what I do. I may mention other methods in passing, but mostly this is just what I currently do. I have done many things in the past and most of them worked more or less. This method is the one I find most reliable in getting cells started, finished and mated. This is not to say that other methods don't work. A lot of what works and doesn't has to do with circumstances. If there is a flow you don't seem to be able to do anything wrong. If there is a dearth things become quite difficult and sometimes you don't seem to be able to do anything right. The reason I do this method is that it has the most success in my experience under adverse conditions as well as good conditions.

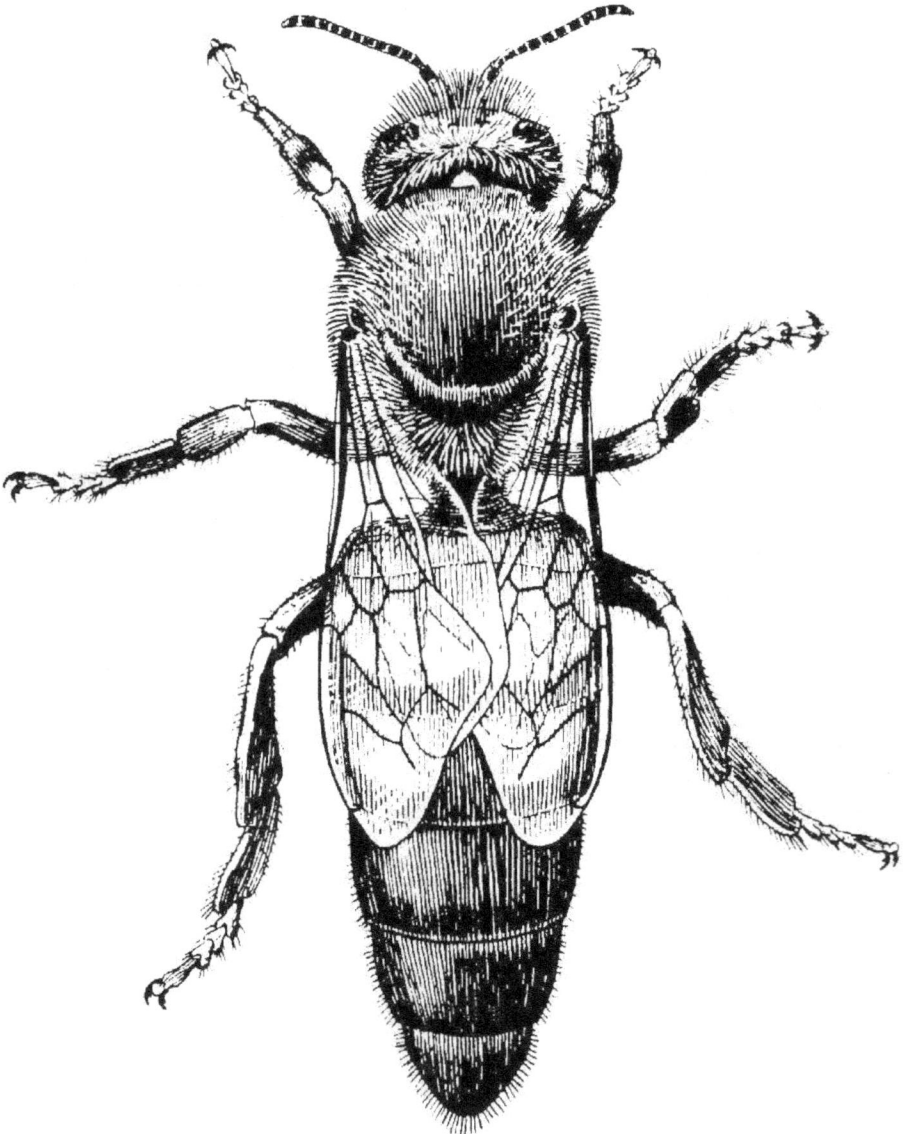

Queen bee

Section 1: Concepts

Queen being balled

Breeding

Let's talk about genetics for a minute. Decisions you make along the way have to do with this and one of the most important reasons to raise your own queens is to get the queens you want and genetics is a third of that equation. The other two thirds are well fed larvae and well bred (meaning a lot of drones) queens. So what do we want in a queen? A good place to start is with feral stock. We want bees with good instincts and the vigor of wildness.

Feral Bees

There is much talk that the feral bees died. In my observation there was a serious shift in what I found when catching feral bees. I used to find "leather" colored Italian looking bees. Now I'm finding more black bees with a little brown mixed in. I'm breeding these survivor bees for myself and for sale

Typically I'm asked how I know these are feral survivors instead of recent escapees. First, they act differently than any of the domestics. Just little things, mostly, but also they overwinter in very small clusters and are very frugal. They are also very variable in aspects that are usually bred out, like propolis use or running on the combs. Also they are typically smaller when you find them, being from natural sized comb.

Swarms

...are the easiest way to get feral bees. But a lot of swarms are, and a lot of swarms aren't, feral bees. I'd take them either way, but if you're looking for feral survivor bees to raise queens then look for the smaller bees. Swarms with small bees are probably feral survivors. Swarms with larger bees are probably swarms from someone's hive. To get swarms, notify the local police and rescue people and the county agricultural extension office. Even the local pest exterminators. If you want to do a lot of them run a yellow pages ad for swarm removal.

Swarm in a tree

Capturing a swarm

Much has been written and each situation is both similar and unique. A swarm is a bunch of homeless bees with a queen. They may have already decided where they think they want to go, or they may still have scouts out looking. Swarms usually happen in the morning and they usually leave by early afternoon, but they may swarm in the afternoon and they may leave in a few minutes or a few days. If you chase swarms you will often get there too late and often get there in time. Both will happen. It's best to have all your equipment with you all the time. If you have to go get your equipment, you will probably be too late. Have a box with a screened bottom attached. This can be attached by nail little squares of plywood into both the box and the bottom or with the 2" wide staples that are sold by bee suppliers for moving hives. You need a lid. I like a migratory cover because it's simple. Less moving parts. I like to have a #8 screen cut and bent to 90 degrees to block the door (but not attached yet). A stapler is nice for anchoring the screen to the door and the cover to the hive. The best are the ones labeled as light duty staplers instead of the heavy duty ones. They penetrate better and stay better. I don't know why. The ones that take the T50 staples are *not* the right ones, although if you already have one you can use it. The ones that take the J21 staples are easier to use. You need a veil minimum, but I like a jacket or a suit. Gloves and a brush are helpful. You can make or buy a rig with a 5 gallon bucket to knock them down with. The idea is that you add EMT (conduit) to it as a long handle and you slam it under the swarm to dislodge them into the bucket. Then you pull on the rope to put the lid on and lower the whole thing back down and dump them on a bedsheet in front of the box. Then if you start tapping (usually called drumming) on the box the bees usually move in. This generally works better than dumping them in the box. If they go in of their own free will they tend to stay. The main trick to swarms is to get the queen. If you can reach and see, try to find the queen. If you know you see her and can make sure she ends up in the box, close it up, brush off the stragglers and leave. If you're not sure, then let them settle in. It helps if the box smells like lemongrass essential oil. Either put some lemongrass essential oil in it (lasts longer) some swarm lure (costs more but works well) or actually spray some lemon pledge (cheap, easy to find, but doesn't last as long) in the box before you put the swarm in. If you pay attention when you buy a package or hive a swarm you'll notice it's what they smell like. Sometimes they will settle into the box. Sometimes you didn't get the queen, or she likes the branch she was on better, and they all start accumulating on the branch again. I just keep shaking them down in front of the hive until they move in. It usually works. In my observation, honey, brood etc. are no help in hiving a swarm although they may help anchor them once they have decided to move in. They are not looking for an occupied house, they are looking for an empty or abandoned house. Old empty comb sometimes helps. Some brood might help anchor them so they don't leave though. It's also well worth having some Queen Mandibular Pheromone. You can either keep your old retired queens in a jar of alcohol (queen juice), or buy PsuedoQueen (last I checked available from Mann Lake).

Swarm shaken onto a bedsheet

Always wear protective equipment. Swarms don't usually get mean, but can be unpredictable. Also be careful of power lines and falling off of ladders. It sounds redundant, but when a lot of bees are buzzing you, and especially if one gets in your bonnet, it's hard to stay calm, but it is a requirement if you are on top of a ladder.

My current favorite method for getting a swarm is skip ladders altogether. Take enough boxes to make a good size (one deep, two mediums) and preferable ones that have been lived in. Some old comb if you have it. Some QMP (a quarter of a stick of PseudoQueen or the end of a Q-tip dipped in queen juice, which is the alcohol from a jar of retired queens in alcohol) and some lemongrass essential oil. Dip the other end of the Q-tip in the lemongrass oil. Drop the Q-tip in the hive, put the lid on, put it near the swarm and come back after dark. They probably will have moved into the box. Staple screen over the entrance and take them home.

Removal

Sometimes called a "cut out". This is not the easiest way to get bees. It is exciting and fun, but sometimes requires some construction skills and lots of courage. The idea is to remove all of the bees and all of the combs from a tree, a house, or whatever they are living in. It often involves removing sections of walls and someone repairing them afterwards. It is not usually financially worth it unless you are being paid to remove them or you have a lot of free time.

Each removal is a separate situation. Sometimes they are in an old abandoned building and the owner doesn't care if you rip the wallboard off or tear the siding off. Usually it does matter and you can't go tearing it up, you have to put everything back when you are done or make it clear to the homeowner that they will need to hire a carpenter to do so. Ignoring, for the moment, the construction issues, if you get to the combs, whether they are in a house or a tree or what-ever, you need to cut the brood to fit frames and tie around the frames to hold it in. This does not work well for honey, especially in new comb, because it's too heavy, also brood has to breath and trying to save the honey usually results in the cappings on the brood getting honey on them and then the pupae suffocate and then the small hive beetles take over, so scrap the honey. Throw it in a five gallon bucket with a lid to keep out the bees trying to clean up the spill. Try to put the brood in an empty hive box and keep brushing or shak-ing the bees off into it. If you see the queen, then catch her with a hair clip queen catcher or put her in a cage and put her in the hive box. If you get some brood and the queen in the hive box the rest of the bees will even-tually follow. If you don't see the queen, then just keep putting bees in the box and brood comb in frames in the box and honey in the bucket until the combs are all gone. Take the bucket and, if you can, leave for a few hours and let the bees figure out where the queen and the other bees are. The will all settle into the new box. At dark they should all be inside and you can close it up and take it home. Some lemongrass oil and some QMP are helpful to lure them into the box and get them to settle there instead of all going back to the wall.

Cone Method

This method is used when it's impractical to tear into a hive and remove the comb or there are so many bees you don't want to face them all at once. This is a method where a screen wire cone is placed over the main entrance of the current home of the bees. All other entrances are blocked with screen wire stapled over them. Make the end of the cone so it has some frayed wires so that a bee can push the wires enough to get out (including drones and queens) but can't get back in. Aim it a bit up and it helps some on keeping them from finding the entrance. Now you put a hive that has just a frame of open brood, a couple of frames of emerging brood and some honey/pollen, right next to the hive. You may need to build a stand or something to get it close to where the returning foragers are clustered on the cone. Sometimes they will move into the box with the brood comb. Sometimes they just hang on the cone. The biggest problem I've had is that this causes many more bees to be looking for a way in and circling in the air and the homeowners often get antsy and spray the bees with insecticide because they are afraid of them. If you think this is likely, then *don't* put the box with the brood here, but rather at your bee-yard, hopefully at least 2 miles away, and you vacuum or brush the bees off into a box every night and take them and dump them in the box with the brood, you will eventually depopulate the hive. If you keep it up until no substantial number of bees are in it anymore, you can use some sulfur in a smoker to kill the bees (sulfur smoke is fatal but does not leave a poisonous residue) or some bee quick to drive the rest of them out of the tree (or house or whatever). And if you use the BeeQuick you may even get the queen to come out. If you do, catch her with a hair clip queen catcher and put her in a box and let the bees move into the box. Since the cone is still on the entrance they can't get back in the old hive. I'd leave it like this for a few days and then bring a strong hive and put it close to the old hive. Remove the cone and put some honey on the entrance to entice the bees to rob it. This is most effective during a dearth. Mid-summer and late fall being likely dearths. Once they start robbing it, they will rob the entire hive out. This is especially important if removing them from a house, so that the wax doesn't melt and honey go everywhere or the honey attract mice and other . Now you can seal it up as best you can. The expanding polyurethane foam you buy in a can at the hardware store is not too bad for sealing the opening. It will go in and expand and make a fairly good barrier. Joe Waggle came up with this option, if you can keep a good eye on it is when they swarm, put the cone on and then the virgin queen will leave to mate and not be able to get back in. Then you can get a swarm with a queen from the cone.

Bee Vacuum

I will preface this that I don't like Bee vacuums. They kill a lot of bees make it hard to find the queen and likely to kill her. I hardly ever use them. They are nice for cleaning up the last stragglers at a colony, but I prefer to use a spray bottle of water to keep them from flying so much and a bee brush or shaking to get them off. I think a Bee vacuum is often a replacement for finesse and skill. Since they are occasionally useful, let's talk about them.

Some of the bee supply houses make these, but you can modify a cheap shop vac to do it. The most important issues are these:

If you have too much vacuum it will kill too many bees. If you are converting a shop vac, cut a hole in the top or use a hole saw and drill a hole. You'll have to adjust this to fit the way the vac is designed, but if there is room you could just drill a three inch hole. If not you could drill and saw to make a longer hole. The idea is that we will take a piece of wood or plastic and make a damper by putting a screw through it on one corner and pivoting the damper to make a larger or smaller hole. This hole is covered on the inside by hardware cloth or screen wire. I just glue it with epoxy on the inside. Now when you adjust the damper to be more open there is less vacuum. When you close it more, there is more vacuum.

If the bees hit the bottom of the vacuum too hard they will die or be injured. The solution to that is put a piece of foam rubber on the bottom. Or wad up some newspaper and put it on the bottom—anything to soften their landing so they don't hit the hard plastic bottom.

Bees get torn up hitting the corrugations of the tube. If you get a smooth hose there will be less of this. If you get smaller corrugations there will be less of this.

If you run the vacuum too long the bees inside get hot, regurgitate their honey and die. If this happens you will notice they are a sticky mess. Don't run the vacuum any longer than you have to.

Adjust the vacuum carefully. You want just enough vacuum to pick the bees off the comb and no more. Too much and you'll have a canister full of squashed bees.

This tool can be used for bee removal. Getting bees off of the combs and not in the air is very helpful. Be careful. I have used them with good luck and I have also killed a lot of bees when I didn't mean to.

Transplanting Bees

Moving bees from one "hive" to another (trees, old hives or other homes of bees).

People often have bees in an old rotting hive that is crumbling to pieces and is so cross-combed they can't manipulate it. Or they have a hive in a log gum, a box hive (no frames), a skep, a piece of a tree that fell down or some oddball equipment that they want to retire or even that they want to move them from all deeps to all mediums etc. If you want bees to abandon some current abode that can be taken home and manipulated here are some methods that I've used, and some variations that I have not used, but should work.

I have used this on box hives and log gums. You want the bees to abandon their old home, but you don't want to sacrifice all the brood. You want to get most of the bees and the queen out of the old hive into a box that is connected to the old hive. In other words there needs to be some connection between the two. A piece of plywood that is as large as the largest dimension of either one in both directions can then had a hole cut in the middle of it that is as large as the smaller of either on in both directions. By putting this between the new hive body and the old hive you have connected the two.

The next decision is whether you want to use Bee Go, Bee Quick (similar but smells nicer) or smoke and drumming or just patience.

It helps if the new hive has some drawn comb and, better yet, a frame of brood.

If you want to use the fumes (Bee Go and Bee Quick) then you put the old hive on top and the new hive body on the bottom. Have a queen excluder handy. Use a soaked rag for fumes and put it as near the top of the old hive as you can. This will drive the bees down into the box. When the box seems pretty full and the old hive seems pretty empty put the excluder between. If you can easily do it, put the old hive so that the combs are upside down from what they used to be. That way the bees will be more likely to abandon it because the cells are sloped down instead of up.

If you want to smoke and drum, then you put the old hive on the bottom and the new one on the top. Smoke the old hive heavily and tap on the side with a pocketknife or a stick. You don't have to do it hard like a bass drum, just a tap tap tap. Lots of smoke helps. Again, when it looks like most of the bees are in the top put in the excluder. It doesn't matter what the orientation of the combs is for driving the bees out, but it helps if it is upside down now. The queen should be in the top and they will finish the brood in the bottom and then rework it for honey or abandon it.

If you want to use patience, just put the new hive on the top and wait for the bees to move up. This may or may not work for some time because the queen wants to stay in the brood chamber.

Bait hives.

Bait hives are empty boxes that are set out to try to entice a swarm to move in. They will not entice a hive to swarm, but they may offer a nice place for a hive that wants to swarm. I use Lemongrass oil and sometimes queen pheromone. You can buy QMP (Queen Mandibular Pheromone brand name PseudoQueen). It is little tubular pieces of plastic that have the smell impregnated in them. When I use these for bait, I cut each of them into four equal pieces and use one piece and some lemongrass oil or some swarm lure. Swarm lure and QMP are available from bee supply places. You can get your own QMP by putting all your old queens when you requeen and any unused virgin queens in a jar of alcohol. Put a few drops of this in the bait hive. Old empty combs are nice too and using boxes that have had bees in them helps. I set out about seven of these last year and got one swarm. Not great odds, but I got some nice feral bees. Once you catch a swarm, I would put more bait hives there and once you haven't gotten a swarm in a year at a location, I would move those to a new location. There are things that have been researched to increase your odds such as the size of the box, the size of the opening and the height in the tree. There seem to be a lot of exceptions though. So far my best luck has been a box the size of a deep five frame nuc or a 8 frame medium with some kind of lure (homemade or otherwise), 12 feet or so up a tree, with about the equivalent of a 1" hole for an entrance. And foundationless frames (frames with a comb guide, see chapter *Foundationless Frames*). My problems have been wasps moving in, finches moving in and wax moths eating old combs and kids knocking them out of trees with rocks and destroying them. Try putting nails in the hole to make an "X" to make it hard for the finches or cover the hole with #4 hardware cloth. Paint them brown or "tree" colored to make them harder for the kids to see. Use starter strips or clean dry old comb so the wax moths don't move in or spray the old comb with Xentari Bt. Remember, this is like fishing. I would not count on it if you're trying to get started beekeeping. You might catch one the first year or you might not catch one for several years, or you might catch several. It's like fishing because you want fish for supper. You may be better off to buy some fish.

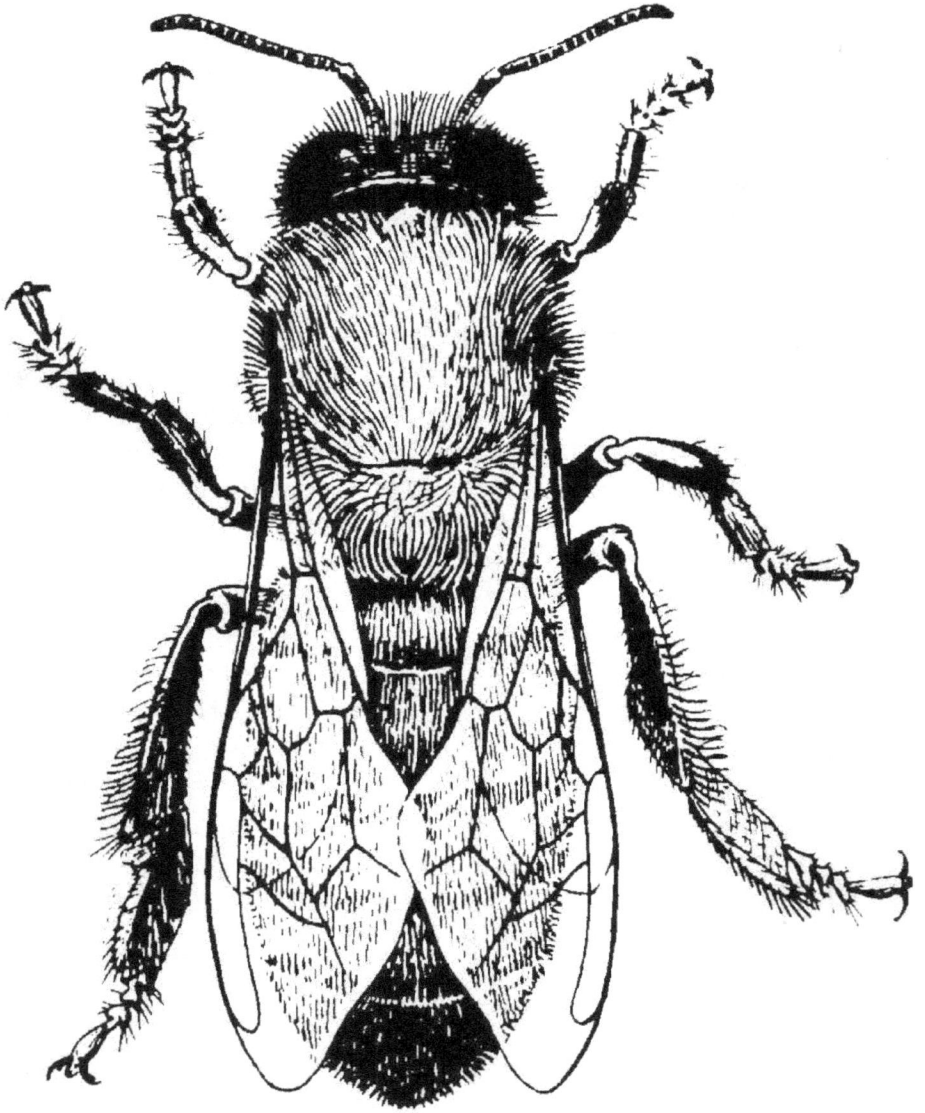

Worker bee

The Whole Bee Concept

Maintaining Genetic Diversity and Locally Adapted Bees

Even if we start with feral stock we still need to select from the stock we have. We also need to do what we can to maintain a broad gene pool.

The danger of breeding for specific traits

The history of selective breeding is full of both successes and failures. Many a great breed became great when the overall health and usefulness was the criteria for selection. And many of those wonderful breeds were ruined when some specific trait became the "trait de jour". I hardly think it necessary to give a lot of examples of this since they are abundant in every domestic species. Dogs, cattle, horses, etc. have all suffered from this mentality of breeders. Let's pick one that seemed practical at the time. Hereford cattle were bred for many years to be "compact". The thinking was that long legs were a waste of energy since you can't sell bone, just meat. So if cattle had short legs with more meat on them and less bone, the animal would be a higher proportion of meat and lower proportion of bone and therefore be more profitable. So for a century or more they were bred to be "compact". The breeding for compactness was a great success if you measure it by just that trait. The problem was when they succeeded the once hardy and self-sufficient cattle breed was suddenly beset by calving difficulties. Someone, shortly after, started correlating leg length and calving difficulties and discovered that short legged cattle had more problems and long legged cattle had less. So now they discovered, after throwing away all of the long legged genetics, that they had backed themselves into a genetic corner. What they should have been breeding for was overall health (including ease of calving).

The appeal of selecting for very specific traits is that it seems so scientific. The problem is that it is not so scientific. Reality is that the genetic combinations that produce health, longevity, productivity etc. are not just one gene or one simple trait, they are a combination of many genes and many traits. The problem of breeding for specific traits is that you are not only missing the "forest for the trees" but you are missing the "forest for the" cells in the leaves on the trees. In other words you need to back up and get some perspective.

Danger of being too selective

"We're trying to ensure the failure of modern beekeeping by focusing too much on single traits; by ignoring the elements of Wildness; and by constantly treating the bees. The biggest mistake of all is to continue viewing mites and other "pests" as enemies that must be destroyed, instead of allies and teachers that are trying to show us a path to a better future. The more virulent a parasite is, the more powerful a tool it can be for improving stocks and practice in the future. All the boring and soul-destroying work of counting mites on sticky boards, killing brood with liquid nitrogen, watching bees groom each other, and measuring brood hormone levels—all done in thousands of replica-

tions—will someday be seen as a colossal waste of time when we finally learn to let the Varroa mites do these things for us. My own methods of propagating, selecting and breeding bees, worked out through many years of trial and error, are really just an attempt to establish and utilize Horizontal breeding with honeybees—to create a productive system that preserves and enhances the elements of Wildness. My results are not perfect, but they have enabled me to continue making a living from bees without much stress, and have a positive outlook for the future. I have no doubt that many other beekeepers could easily achieve these same results, and then surpass them."—Kirk Webster, What's missing from the current discussion and work related to bees that's preventing us from making good progress. kirkwebster.com

The other issue of being too selective is that you can create a bottleneck in the gene pool. Genetic bottlenecks can cause obscure problems to become common problems. Inbreeding fixes traits. The problem is that it fixes both good and bad traits. Fixing the bad traits can result in making these traits endemic to the population. Genetic bottlenecks also can eliminate lines that might be needed to survive the next "bee crisis".

A catastrophic example of this was the Irish potato famine. First of all there are thousands of varieties of potato and only a few are susceptible to the blight that caused the famine. Second, the potatoes were started by cutting eyes, which means that the potatoes were all clones of each other, with no genetic diversity from one potato to the next. So this genetic bottleneck led to the deaths of over a million people and the displacement of millions more. This is the danger of genetic bottlenecks.

Complexity of Success

The genetic combinations that lead to success are almost infinitely complex. The combinational analysis of what makes a gentle, productive, healthy bee is beyond our comprehension. But observing success is not beyond our comprehension.

Success may not even be genetic

The success of a hive is so complex that it may be we are actually choosing based on the microbe's genetics rather than the bees' or even misinterpreting success altogether.

Jay Smith shares this story in "Better Queens":

"In Indiana we had an outyard laid out in the form of a triangle as that was the shape of the plot on which we had our bees. During the sweet clover flow one colony produced three supers of honey while the others averaged about two supers. In the fall that colony produced two supers of honey from smartweed and asters while the rest produced a little less than one super. Surely that colony that so far outdistanced the others must have a queen that would make an excellent breeder.

"I thought I would take a look at her but alas, when I opened the hive, I found it not only had no queen but was fairly lousy with laying workers! Just why then the big yield? This colony was located at the point of the triangle to the west and the fields of nectar lay to the west. It was evident that the bees in returning from the fields-maybe the

ones out for their first load-stopped at the first hive they came to and kept it packed with bees."

What we have bred for
- No Propolis
- Solid brood patterns
- Queens that never shut down
- Color
- Large bees
- Less drones
- Less swarming
- More honey

Counter Productive

We have bred bees that are not as healthy because propolis is part of their immune system.

We bred against hygienic behavior by breeding for perfect brood patterns.

We bred for longer gestation times (giving the Varroa an advantage) by breeding for larger bees.

We have bred bees that are reproductively challenged because of less drones, bigger drones, less swarming etc. giving the edge to the Africanized honey bees or other wild bees

Basically almost everything we bred for was a bad idea.

What we should breed for
- Overall health and vitality
- Ability to detect a failing queen and replace her
- Adapted to your climate and your flows
- Productivity
- Overwintering

- Gentle and manageable

How do you assess?
- You need them to at least have gone through one winter with that queen's workers.
- You need them to at least have gone through one flow with that queen's workers.
- Look at the big picture of health and good instincts. Not single traits.

Maintaining genetic diversity
- Don't breed all your queens from the same line
- Think more in terms of removing what you don't want.
- Breed out the bad.
- Leave all of the good ones and try to maintain all of the lines that are worth keeping.
- The goal is for the gene pool to be both broad and good.

Breeder queen in at least her 2nd year
Some of the great queen breeders such as Jay Smith had breeder queens that were 6 or 7 years old How can you assess a queen if you haven't seen her offspring overwinter successfully and produce well?

Bees with a gambling problem
Bees are all gamblers. They have to rear brood ahead of the flow to have foragers for the flow. The ones that gamble big are the ones the win big. The ones that gamble big are also the ones that lose big. One theory is that you should breed from "average" bees instead of the "outliers" to avoid the big gamblers.

Maybe we make it too complicated

"The records are carefully scanned, and that queen chosen which, all things considered, appears to be the best. The first point to be weighed is the amount of honey that has been stored. Other things being equal, the queen whose workers have shown themselves the best stores will have the preference.

"The matter of wintering will pretty much take care of itself, for a colony that has wintered poorly is not likely to do very heavy work in the harvest. The more a colony has done in the way of making preparations for swarming, the lower will be its standing. Generally, however, a colony that gives the largest number of sections is one that never dreamed of swarming.

"I am well aware that I will be told by some that I am choosing freak queens from which to rear; and that it would be much better to select a queen whose royal daughters showed uniform results only a little above the average. I don't know enough to know whether that is true or not, but I know that some excellent results have been obtained by breeders of other animals by breeding from sires or dams so exceptional in character that they might be called freaks.

"I know, too, that it is easier to decide which colony does best work than it is to decide which queen produces royal progeny the most nearly uniform in character. By the first way, too, a queen can be used a year sooner than by the second way, and a year in the life of a queen is a good deal. I may mention that a queen which has a fine record for two successive seasons is preferred to one with the

*same kind of a record for only one season."--C.C.
Miller, Fifty Years Among the Bees*

Brood Comb

Queen Rearing Concepts

Here I will talk about some of the concepts of queen rearing and in the process will mention some methods that I won't necessarily be using just so you will be familiar with them. You can find the books for Doolittle and Miller and Hopkins on my web site or you can buy all of these in a compendium under the title "Classic Queen Rearing Compendium". But here I will just touch on the concepts and some of the timing.

For a live presentation by the author of this try a search for videos on the web for "Michael Bush Queen Rearing".

Why rear your own queens?

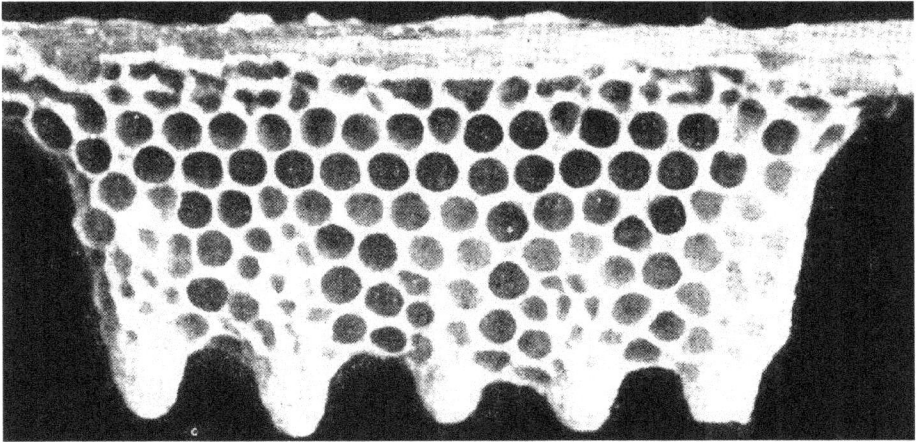

Cost

As of this writing typical queen costs the bee-keeper close to $50 and up counting shipping and may cost considerably more.

Time

In an emergency you order a queen and it takes several days to make arrangements and get the queen. Often you need a queen yesterday. If you have some in mating nucs, on hand, then you already have a queen.

Availability

Often when you need a queen there are none available from suppliers. Again, if you have one on hand availability is not a problem.

AHB

Southern raised queens are more and more from Africanized honey bee areas. In order to keep AHB out of the North we should stop importing queens from those areas.

Acclimatized bees

It's unreasonable to expect bees bred in the deep South to winter well in the far North. Local feral stock is acclimatized to our local climate. Even breeding from commercial stock, you can breed from the ones that winter well in your locale.

Mite and disease resistance

Tracheal mite resistance is an easy trait to breed for. Just don't treat and you'll get resistant bees. Hygienic behavior, which is helpful to avoid AFB (American foulbrood) and other brood diseases as well as Varroa mite problems. And yet most queen breeders are treating their bees and not selecting, either on purpose or by default for these traits. The genetics of our queens if far too important to be left to people who don't have a stake in their success. People selling queens and bees actually make more money selling replacement queens and bees when the bees fail. Now I'm not saying they are purposely trying to raise queens that fail, but I am saying they have no financial incentive to produce queens that don't. This is not to say that some responsible queen breeders aren't doing the right thing here, but most are not. Basically to cash in on the benefits of not treating, you need to be rearing your own queens.

Quality

Nothing is more important to success in beekeeping than the queen. The quality of your queens can often surpass that of a queen breeder. You have the time to spend to do things that a commercial breeder cannot afford to do. For instance, research has shown that a queen that is allowed to lay up until it is 21 days old will be a better queen with better developed ovarioles than one that is banked sooner. A longer wait will help even more, but that first 21 days is much more critical. A commercial queen producer typically looks for eggs at two weeks and if there are any it is banked and eventually shipped. You can let yours develop better by spending more time.

Concepts of Queen Rearing

Reasons for bees to rear queens

Bees rear queens because of one of four conditions:

Emergency

There is suddenly no queen so a new queen is made from some existing worker larvae.

Supersedure

The bees think the queen is failing so they rear a new one.

Reproductive swarming

The bees decide there are enough bees, and enough stores and enough of the season left to cast a swarm that has a good chance of building up enough to survive the winter without endangering the survival of the colony.

Overcrowding swarm

The bees decide that there are too many bees and not enough room or not enough stores to continue under the current conditions, so they cast an over-crowding swarm as population control. This swarm doesn't have the best chance of survival but the colony believes it improves the colony's chances of survival.

We get the most cells and the best feeding for the queens if we simulate both Emergency and Overcrowding.

A beekeeper can easily get a queen simply by making a queenless split with the appropriate aged larvae. So why would we want to do queen rearing?

Most Queens with Least Resources

The underlying concept of queen rearing is to get the most number of queens from the least resources from the genetics chosen for the traits you want.

To illustrate the resource issue let's examine the extremes. If we make a strong hive queenless. They could have, during that 24 days of having no laying queen, reared a full turnover of brood. The queen could have been laying several thousand eggs a day and a strong hive could easily rear those several thousand brood. Then we have lost the potential for about 30,000 or more workers by making this hive queenless and resulted in only one queen. And, actually, this hive made many queen cells, but they were all destroyed by the first queen out.

If we made a small nuc we would only have a couple of thousand queenless bees rearing several queen cells and those couple of thousand bees could only have reared a few hundred workers in that time. But again they made several queen cells and the results were only one queen.

In most queen rearing scenarios we are making the least number of bees queenless for the least amount of time and resulting in the most number of laying queens when we are done.

Where queens come from

A queen is made from a fertilized egg, exactly the same as a worker. It's the feeding that is different and that is only different from the fourth day on. So if you take a newly hatched worker egg, and put it in a queen cell (or in something that fools the bees into thinking it's a queen cell) in a hive that needs a queen (swarming or queenless) they will make those into queens.

Methods of getting larvae into "queen cups"

There are many methods. You can find the original books for many of these here:

http://bushfarms.com/beesoldbooks.htm

Here are a few of them:

The Doolittle Method

Originally published by G.M. Doolittle, is to graft the appropriate aged larvae into some homemade wax cups. This requires a bit of dexterity and good eyesight, but is the most popular method used. Today plastic cups are often used in place of wax. The queen is sometimes confined to get the right aged larvae all in one place for easy selection. #5 hardware cloth works well for this as the workers can pass through it but the queen cannot. This is usually put on old dark brood comb to make the larvae easier to see and to make the cell bottom more sturdy for grafting. Once you have a good eye for the right age larvae this is less critical and one can do this by simply finding the right age larvae. On day 14 these are usually put in mating nucs.

The Jenter method

Several variations of this are on the market under various names. The concept is that the queen lays the eggs in a confinement box that looks like worker cells. Every other cell bottom of every other row has a plug in the bottom. When the eggs hatch the plug is removed and placed in the top of a cup. This accomplished the same thing as the Doolittle method without the need for so much dexterity and eyesight. On day 14 these are usually put in mating nucs.

Jenter Box Front

Jenter Box Back

Jenter Box Top

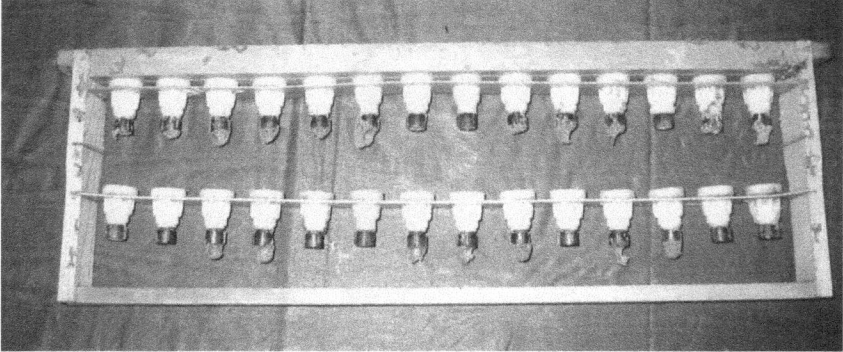

Missed queen cell results in dead queens

Jenter queen rearing system pictures. Front, back and top of the queen box and then a picture of a cell bar where I missed a queen cell the bees built in the cell builder. 17 queens dead.

Advantages to Jenter
• If you are a newbee you get to see exactly what the right age larvae looks like as you know when they were laid.
• If your eyesight isn't so good you don't have to be able to see the larvae (mine isn't the greatest)
• If you are not very coordinated (and I'm not) you don't have to be able to pick up something very tiny and down inside a cell without damaging it. You just move the plugs.

Advantages to grafting
• If the queen didn't lay in the Jenter cage and I'm on a schedule, I don't have any larvae the right age unless I go find some and graft (or do the Better Queens method).
• If I was too busy to confine the queen four days ago, I can just graft.

• If the queen mother is in an outyard, I don't have to make two trips, one to confine her, and another to transfer larvae.
• I don't have to buy a queen rearing kit.

The Hopkins Method

In my variation, the queen is confined with #5 hardware cloth to get her to lay in the new comb and so we know the age of the larvae (as the Doolittle method but on new comb empty instead of old comb). This should be wax, preferably with no wires so you can cut the cells out without wires in the way, although Hopkins says you should used wired comb so it doesn't sag. If you use wired comb, be sure to work around the wires when leaving larvae so the wires won't interfere. Release the queen the next day. You can also just put the new comb in the middle of the brood nest and check every day to see if the queen has laid in it yet, to judge the age of the larvae.

On the fourth day (from when the queen was confined or she laid in the comb) the larvae will be hatched. In every other row of cells *all* the larvae are destroyed by poking them with a blunt nail, a kitchen match head, or similar instrument. Then the larvae in *every other cell* in the remaining rows is destroyed the same way (or two cells destroyed and one left) to leave larvae with space between them. This is then suspended flatways over a queenless hive. A simple spacer is an empty frame under the frame with the cells and a super over that. This will require angling the frames somewhat and laying a piece of cloth on top. The bees perceive these to be queen cells, because of the orientation, and build cells off of them. They should be spaced enough apart to allow cutting them out on day 14 and distributing them to either hives to be requeened (that would have been dequeened the day before) or to mating nucs.

#5 hardware cloth queen confinement cage

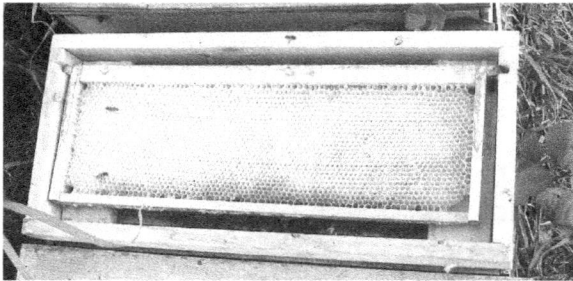

Frame of larvae in the Hopkins shim.

Hopkins shim to hold the frame over the box.

Cell Starter

For me the most difficult thing to get a grasp on and the most critical thing for queen rearing, other than the obvious issues of timing, was the cell starter. The most important thing about a cell starter is that it's overflowing with bees. Queenless is helpful too, but if I had to choose between queenless and overflowing with bees, I'd go for the bees. You want a very high density of bees. This can be in a small box or a large hive, it's the density that is the issue, not the total number. There are many different schemes to end up with queenless crowded bees that want to build cells, but don't ever expect a good amount of cells from a starter that is anything less than overflowing with bees.

The next most important issue with the starter is that it's well fed. If there is no flow you should feed to make sure they feed the larvae well.

Most of the rest of the complexity of the many queen rearing systems, which often seem at odds with one another, are tricks to getting consistent results under all circumstances. In other words, they are important to a queen breeder who needs a consistent supply of queens from early spring until fall regardless of flow and weather. For the amateur queen breeder, these are probably not so important as is the timing of your attempts. Rearing queens during prime swarm season just before or during the flow is quite simple. Rearing queens in a dearth or later or earlier than the prime swarm season will require more "tricks" and more work. For starters I would skip these "additions" and adopt them one at a time as you see the need.

A Cloake board (Floor Without a Floor) is a useful method. You can rearrange things so that part of the hive is queenless during the starter period and queenright as a finisher without a lot of disruption of the hive. But it's not necessary.

The simplest way I know of is to remove a queen from a strong colony the day before and cut it down to minimum space (remove all the empty frames so that you can remove some boxes and, if there are supers that are full remove those). This may even put them in a mood to swarm, but that will make a lot of queen cells. Make sure there aren't any queen cells when you start and if you use them for more than one batch be extra sure there are no extra queen cells in the hive as those will emerge and destroy your next batch of cells.

Another method is to shake a lot of bees into a swarm box aka a starter hive and give them a couple of frames of honey and a couple of frames of pollen and a frame of cells.

Beekeeping Math

Caste	Hatch	Cap	Emerge	
Queen	$3^1/2$	8 +-1	16 +-1	Laying 28 +-5
Worker	$3^1/2$	9 +-1	20 +-1	Foraging 42 +-7
Drone	$3^1/2$	10 +-1	24 +-1	Flying to DCA 38 +-5

Queen Rearing Calendar:

Using the day the egg was laid as 0 (no time has elapsed)

Bold items require action by the beekeeper.

Day Action Concept

-4 Put Jenter cage in hive Let the bees accept it, polish it and cover it with bee smell

0 Confine queen—So the queen will lay eggs of a known age in the Jenter box or the #5 wire cage

1 Release queen—So she doesn't lay too many eggs in each cell, she need to be released after 24 hours

3 Setup cell starter Make them queenless and make sure there is a *very* high density of bees.—This is so they will want queens and so they have a lot of bees to care for them. Also make sure they have plenty of pollen and nectar. Feed the starter for better acceptance.

$3^1/2$ Eggs hatch

4 Transfer larvae and put queen cells in cell starter. Feed the starter for better acceptance.

8 Queen cells capped

13 Setup mating nucs Make up mating nucs, or hives to be requeened—So they will be queen-less and wanting a queen cell. Feed the mating nucs for better acceptance.

14 Transfer queen cells to mating nucs.—On day 14 the cells are at their toughest and in hot weather they may emerge on day 15 so we need them in the mating nucs or the hives to be re-queened if you prefer, so the first queen out doesn't kill the rest.

15-17 Queens emerge (In hot weather, 15 is more likely. In cold weather, 17 is more likely. Typical-ly, 16 is most likely.)

17-21 Queens harden

21-24 Orientation flights

21-28 Mating flights

25-35 Queen starts laying

28 If you intend to requeen your hives, look for laying queens in the mating nucs. If found, dequeen hive to be requeened.

29 Transfer laying queen to queenless hive to be requeened.

Queen marking colors:
Years Ending in:
- 1 or 6 – White
- 2 or 7 – Yellow
- 3 or 8 – Red
- 4 or 9 – Green
- 5 or 0 - Blue

Queen Catching and marking

Until you get the hang of it, there is always the risk of hurting the queen. But learning to do it is a worthwhile undertaking. I would buy a hair clip queen catcher and a marking tube and paint pens. Practice on a few drones with a color from a couple of years ago, or better yet the color for next year, so you don't confuse the drones with the queen. Use the current color for the queens.

My preferred method is to buy a "hair clip" queen catcher, a queen muff (Brushy Mountain) and a marking tube and a marking pen. Catch the queen gently with the hair clip. It is spaced so as not to easily harm the queen, but still be careful. If you put this and the marking tube and the paint pen (after it is shaken and started) in the queen muff then the queen can't fly off while you do this. Take the marking tube and slid out the plunger. If you move away from the hive you can lose some of the bees that are in and on the clip. Don't shake it while holding the clip portion or you may shake the queen out. If you take it in a bathroom with a window and turn off the lights you can be more assured she won't fly off. Or buy a queen muff from Brushy Mountain. Use a brush or a feather and brush off the workers as they come out and then try to guide the queen into the tube. She tends to go up and she tends to go for the light, so open the clip so she will run into the tube. If she doesn't and she runs onto your hand or glove, don't panic, just quickly drop the clip and gently but quickly put the tube over her. Cover the tube with your hand to block the light so she runs to the top of the tube. Put the plunger in. Be quick but don't hurry too much. Gently pin the queen to the top of the marking tube and touch a small dot of paint (start the paint pen on a piece of wood or paper first so there is paint in the tip already) on the middle of the back of her thorax right between her wings. If it doesn't look big enough just leave it. You need to keep her pinned for several more seconds while you blow on the paint to dry it. Don't let her go too soon or the paint will get smeared into the joint between her body sections and it may cripple or kill her. After the paint is dry (20 seconds or so) back the plunger up to halfway so the queen can move. Pull the plunger and aim the open end to the top bars and the queen will usually run right back down into the hive.

Queen longevity:

"In Indiana we had a queen we named Alice which lived to the ripe old age of eight years and two months and did excellent work in her seventh year. There can be no doubt about the authenticity of this statement. We sold her to John Chapel of Oakland City, Indiana, and she was the only queen in his yard with wings clipped. This, however is a rare exception. At the time I was experimenting with artificial combs with wooden cells in which the queen laid."—Jay Smith, Better Queens, Pg 18

I would point out that Jay says: "This, however is a rare exception."

I think three years has always been pretty typical of the useful life of a queen.

Queen Banks

A beekeeper can keep a number of queens in one hive if you get bees that are in the mood to accept a queen (queenless overnight or a mixture of bees shaken from several hives) and the queens are in cages so they can't kill each other. I've done these with a $^3/_4$"

shim on top of a nuc or a frame with plastic bars that hold the JZBZ cages. I put a frame of brood in periodically to keep them from developing laying workers or running out of young bees to feed the queens.

FWOF

(Floor With Out a Floor aka Cloake Board). Used to allow converting a top box on a queen rearing hive to change from a queenless cell starter to a queenright cell builder or finisher. This one is made with a $^3/_4$" by $^3/_4$" piece of wood with a $^3/_8$" x $^3/_8$" groove in it. Hang it out $^3/_4$"or more in front and put a piece across the front under the sides to make a landing board. Cut a piece of $^3/_{16}$" or $^1/_4$" luan to slide in for a removable bottom. Coat edges with Vaseline to keep the bees from gluing it in. From left to right: The frame on a hive with the floor out. Inserting the floor. The FWOF with the floor in. The idea is to put the tray in and face the entrance of the FWOF the direction of the current entrance and face the bottom board in the opposite direction which funnels all the bees to the top box (which goes on top of the FWOF). If you put the queen in the bottom box with an excluder over it you can put open brood and the queen cells in the top box with the floor in for the first 48 hours or so until the cells are started. Then you can pull it out to make it a queenright finisher.

Queen cell bar frame

Hives all in a row.

Queen Rearing Method Names

This information is interesting but not necessary to queen rearing. I've always found the names of queen rearing methods a bit confusing. The more I researched them the more confusing I found them. Here are some of my discoveries.

Queens Rearing Methods and names in chronological order:

Doolittle Method from Scientific Queen Rearing, 1846, Transferring larvae into artificial queen cups.

Nicol Jacobis was a German scientist/beekeeper who published in Die rechte Bienen-Kunst that workers could raise a new queen from a young worker larvae in 1568 and wrote about grafting.

Schirach, based on this work also wrote about grafting. -- A.M. Schirach, ("Physikalische Untersuchung der bisher unbekannten abet nachher entdeckten Erzeugung der Bienenmutter," 1767):

"M. Schirach's famous experiment on the supposed conversion of a common worm into a royal one, cannot be too often repeated, though the Lusatian observers have already done it frequently. I could wish to learn whether, as the discoverer maintains, the experiment will succeed only with worms, three or four days old, and never with simple eggs."— Francis Huber, New Observations on the Natural History of Bees Letter IV.

Which Huber repeated in 1789 and published in 1794:

"I put some pieces of comb, with some workers eggs, in the cells, and of the same kind as those already hatched, into a hive deprived of the queen. The same day several cells were enlarged by the bees, and converted into royal cells, and the worms supplied with a thick bed of jelly. Five were then removed from those cells, and five common worms, which, forty-eight hours before we had seen come from, the egg substituted for them. The bees did not seem aware of the change; they watched over the new worms the same as over those chosen by themselves; they continued enlarging the cells, and closed them at the usual time" –Francis Huber, New Observations on the Natural History of Bees Letter IV.

And Doolittle repeated in 1846:

"In this work I often found partly-built queen-cells with nothing in them, or perhaps some would contain eggs, which, when I found them, I would take out, substituting the larvae in their places. As a rule, I would be successful with these, as well as with those that were put into the cells that contained royal jelly, but now-and-then a case would occur when only those placed in royal jelly would be used."—G.M. Doolittle, Scientific Queen Rearing Chapter V

Doolittle does not take credit for inventing the queen cup:

"I remember that away back in some of the bee-papers, some one had proposed making queen-cells to order, on a stick, for a penny a piece, and why could I not so make them? It would do no harm to try, I thought; therefore I made a stick, so that it would just fit inside of a queen-cell, from which a Queen had hatched, and by warming a piece of wax in my hand, I could mould it around the stick, so as to make a very presentable queen-cup."

So the "Doolittle method" was not, by Doolittle's admission, invented by Doolittle. This should be the "Schirach Method" or even more accurately, the "Jacobis Method".

Alley Method from The Bee-Keeper's Handy Book I, 1883, Starting in a swarm box and cutting worker comb and attaching vertically instead of horizontally

Alley only used the "swarm box" as a way to convince the bees of their queenlessness. Then he put them in a queenless cell starter hive. So the concept of starting cells in a "swarm box" did not originate with Alley since he never actually used it in the current way it is used.

Alley suggested using old brood comb and he attached it to the bottom of existing comb and not a "cell bar".

Elements of this show up in the Hopkins and Smith methods, but no one, that I know of, is using the Alley method commercially though parts of it show up in other methods.

Miller Method from A Year Among the Bees to Fifty Years Among the Bees, 1885, Cutting a new comb in a zig zag or a foundation in a zig zag and strips.

Albert Cook published this in 1876. As far as I know Miller never claimed he originated this method, he just popularized it. So the "Miller" method is really the "Cook" method. There are still hobbyists doing this. I know of no commercial queen breeders doing this.

Hopkins Method aka Case Method, Isaac Hopkins from the Australasian Bee Manual, 1911, Turning a comb of worker brood horizontal with larvae destroyed to make gaps between the queen cells

But Isaac Hopkins gives credit to an unnamed Austrian beekeeper for inventing the method that is usually attributed to him. This method is in the 1911 version of his book.

His own method was a modified "Alley method" with new comb instead of old comb waxed to cell bars, instead of the bottom edge of some comb. About his own method Hopkins says:

> *"I have tried several methods for raising queen cells, but none have given me so much satisfaction as the one I first saw described in Gleanings in Bee Culture for August, 1880 by Jos. M. Brooks and which I have since practiced. It is very similar to Mr. Alley's method, explained in his "Handy Book" "—Isaac Hopkins, The Australasian Bee Manual 1886 Chapter XII pg 211*

So Hopkins actually used a modified Alley Method, basically substituting new comb for old comb which he lays no claim to, but instead gives credit to Joseph Brooks.

I can't find information on what "Case" recommended, so I can't say if this is really the "Case Method" or not.

Smith Method from Queen Rearing Simplified, 1923, Starting cells in a "swarm box" and grafting larvae into cups.

Smith gets credit for originating starting the cells in the swarm box, rather than just using it to convince the bees of their queenlessness as Alley did. Smith, however gives that credit to Eugene Pratt. Grafting, of course, is the "Doolittle method" which was invented by Schriach or Jacobis. Usually when referring to the "Smith Method" as opposed to the "Doolittle Method" the distinction is in the use of the "Swarm Box".

Smith's Better Queens Method 1949, Better Queens, Cutting strips of new comb with worker brood and destroying every other cell and putting it on a cell bar

This, of course, is the actual Hopkins method (or more accurately the Joseph M. Brooks method), which Hopkins wrote about 63 years before Jay Smith did. Which is really only the Alley method with new comb. I have no doubt Smith came up with it himself after observing emergency cells on new comb compared to emergency cells on old comb, but the main concept is a rehash of the Alley method with new comb and a cell bar. Of course there are many details that Smith had refined over the years, but the basic concept, as Jay Smith himself says, is just the Alley method with new comb.

If it deserved a new name to distinguish it from the "Alley Method" this should be the "Hopkins Method" or, better yet, the "Brooks Method" or maybe just the "Alley Method with new comb".

OTS aka Disselkoen method tearing down cell walls below the larvae you want to use

Also called the "On The Spot" queen rearing method. Probably, just like Jay Smith in his Better Queens method, Mel probably didn't know this had already been written about. Of course, just because someone writes about it doesn't mean they invented it. But this had been written about before.

"... queens may be developed through the power of suggestion, as follows: Select a frame of brood from the best colony; with a toothpick tear down the partitions between three worker cells which contain eggs or larvae less than two days old and destroy two of the eggs or larvae; repeat the oper-ation in several places.

Place the frame back in the hive, being very sure that there is sufficient space between it and its neighbouring frame, so that good queen cells may be built. If there is a scarcity of honey, feed the bees. The cleverness of bees is clearly proven by the readiness with which they take a hint, and they almost invariably build queen cells upon the comb thus treated."-- How to Keep Bees (A Handbook for the Use of Beginners)' by Anna Botsford-Comstock, 1905, pp.156-7

Of course it's too late to straighten it all out now.

A Few Good Queens

Before I get too far into my queen rearing methods, what if you only want a few queens? Here is a simple method for that and again we will point out some of the concepts.

Simple Queen Rearing for a Hobbyist

I get this question a lot, so let's simplify this as much as possible while maximizing the quality of the queens as much as possible.

Labor and Resources

The quality of a queen is directly related to how well she is fed which is related to the labor force available to feed the larvae (density of the bees) and available food.

Equipment

Second let's talk about equipment. One can set up mating nucs in standard boxes with dummy boards (or division boards) but only if you have the extra boxes or division boards. The advantage is that you can expand this as the hive grows if you don't use the queen. You can also build either two frame boxes or divide larger boxes into two frame boxes (commonly sold as queen castles). These need to be the same depth as your brood frames.

Method:

Make sure they are well fed

Feed them for a few days before you start unless there is a strong flow on.

Make them Queenless

So if we make a hive queenless (do what you like about having new comb or not) nine days after making them queenless these will be mostly mature and capped and be three days from emerging.

Make them Crowded

You actually do this at the same time you make them queenless. Go through the hive looking for the queen while also removing any empty combs. What you are doing is what I call "compressing the hive". You want to take all the empty combs and boxes of empty combs off of the hive. If the bees still all fit in the colony, then take a box and shake all the bees out and give that box to another hive. Keep compressing them until the hive is overflowing with bees.

Make up Mating Nucs

Now at 9 days from when we made them queen-less we need somewhere to put the cells. Unless you intend to use the cells to requeen your hives directly, we need to set up mating nucs. The "queen castles" or four way boxes that take your standard brood frames and make up four, two frame mating nucs in one box are very good for this, but dummy boards and regular boxes can work also. In my operation these are all medium depth two frame nucs. The queen we removed earlier goes well in one of these also. We now want a frame of brood and a frame of honey in each of the mating nucs.

Transfer Queen Cells

The next day (ten days after making them queen-less) we will cut out (with a sharp knife) the queen cells from the new wax combs we put in. If we used unwired foundation (or none) they should be easy to cut out without running into obstacles (as we would with wire and with plastic foundation) and can put each of the cells in a mating nuc. You can just press an indentation with your thumb and gently place the cell in the indentation. If you want you can also just put each frame that has cells on it in a mating nuc and sacrifice the extra cells (as the first queen out will destroy them). This is helpful if you have plastic foundation or you just don't want to mess with cutting out cells.

Check for Eggs

Two weeks later we should see some eggs in the mating nucs. If not, then by three weeks we should. Let her lay up the nuc well before moving her to a hive or caging her and banking her for later.

Next round just make the mating nuc queenless again the day before adding cells.

Now that these nucs are well populated by the brood the queen has laid, we can make more queens by simply making a strong mating nuc queenless and they will raise more queens. Again, it's the density of bees and the supply of food that are the issues. We can also, if they are wax combs, cut cells out and make use of multiple cells in other mating nucs as well. In this case either set up those nucs the day before or remove the queen the day before.

And that is all there is to raising a few queens.

Entrance

Quality of Emergency Queens

One of the controversies in beekeeping is the quality of emergency queens. The previous chapter would result in emergency queens and some people believe that the bees will start with too old of a larvae. So let's discuss this.

The experts on emergency queens:

Jay Smith, from Better Queens

"It has been stated by a number of beekeepers who should know better (including myself) that the bees are in such a hurry to rear a queen that they choose larvae too old for best results. later observation has shown the fallacy of this statement and has convinced me that bees do the very best that can be done under existing circumstances.

"The inferior queens caused by using the emergency method is because the bees cannot tear down the tough cells in the old combs lined with cocoons. The result is that the bees fill the worker cells with bee milk floating the larvae out the opening of the cells, then they build a little queen cell pointing downward. The larvae cannot eat the bee milk back in the bottom of the cells with the result that they are not well fed. However, if the colony is strong in bees, are well fed and have new combs, they can rear the best of queens. And please note-

- they will never make such a blunder as choosing larvae too old."--Jay Smith

Quinby seems to agree:

"I want new comb for brood, as cells can be worked over out of that, better than from old and tough. New comb must be carefully handled. If none but old comb is to be had, cut the cells down to one fourth inch in depth. The knife must be sharp to leave it smooth and not tear it."--Moses Quinby

C.C. Miller's view of emergency queens

"If it were true, as formerly believed, that queenless bees are in such haste to rear a queen that they will select a larva too old for the purpose, then it would hardly do to wait even nine days. A queen is matured in fifteen days from the time the egg is laid, and is fed throughout her larval lifetime on the same food that is given to a worker-larva during the first three days of its larval existence. So a worker-larva more than three days old, or more than six days from the laying of the egg would be too old for a good queen. If, now, the bees should select a larva more than three days old, the queen would emerge in less than nine days. I think no one has ever known this to occur. Bees do not prefer too old larvae. As a matter of fact bees do not use such poor judgment as to select larvae too old when larvae sufficiently young are present, as I have proven by direct experiment

and many observations."--Fifty Years Among the Bees, C.C. Miller

---

A study:

David C. Gilley, David R. Tarpy, Benjamin B. Land: Effect of queen quality on interactions between workers and dueling queens in honeybee (Apis mellifera L.) colonies

---

Selection of high-quality queens by the workers during queen development has been demonstrated by Hatch et al. (1999), who found that during emergency queen rearing (the process by which workers rear queens from worker larvae to replace a queen that has died unexpectedly) workers preferentially destroyed queen cells built from older worker larvae. Despite selective behavior by the workers during queen rearing, considerable variation in quality exists among newly emerged adult queens (Eckert 1934; Clarke 1989; Fischer and Maul 1991). This variation in quality among queens gives workers the opportunity to benefit by selecting high quality queens that are fully developed, when the decision will be most accurate.

My Take

The concepts of getting a good queen are to get the right age larvae, to get it well fed, to get the resulting queen well bred, and to allow that queen to lay long enough for good ovariole development.

Right Age

I agree with the above experts that the bees won't raise a queen from too old of a larvae. If they start a cell from too old of a larvae, I believe they will tear it down anyway. But if you wish to stack the deck in this regard, tearing down the cell wall (per Mel Disselkoen) or using new comb (per Smith and Quinby) or cutting the bottom edge of the comb (Miller) can be done.

This idea has been around a while:

"... the time when queen cells are naturally built may not be the most convenient or the most desirable time for giving certain colonies a new mother. This being the case, queens may be developed through the power of suggestion, as follows: Select a frame of brood from the best colony; with a toothpick tear down the partitions between three worker cells which contain eggs or larvae less than two days old and destroy two of the eggs or larvae; repeat the operation in several places.
Place the frame back in the hive, being very sure that there is sufficient space between it and its neighbouring frame, so that good queen cells may be built. If there is a scarcity of honey, feed the bees. The cleverness of bees is clearly proven by the readiness with which they take a hint, and they almost invariably build queen cells upon the comb thus treated."-- How to Keep Bees (A Handbook for the Use of Beginners)' by Anna Botsford-Comstock, 1905, pp.156-7

Well Fed

Accomplishing this basically takes two things: a lot of resources coming in (not a dearth or feeding if there is) and a high density of bees. One way to accomplish this is simply to do it during the early spring flow when a lot of resources are readily available. Another is to feed pollen and syrup or honey during the days before the queen cells are capped. In a dearth you will probably also need to feed the mating nucs to get the queens to fly out to mate. This can be done in the evening in small amounts so it is cleaned up by morning and doesn't start robbing. I usually use a small conical dixie cup squashed to fit between the frames with a couple of tablespoons of syrup in it.

Well Bred

Accomplishing this basically is simply to do it during the early spring when a lot of drones are flying.

Ovariole development

All you have to do is let her lay for at least two weeks once she starts. If this is in a mating nuc, basically wait until you have a lot of capped brood and you'll probably be about right.

How to do it wrong

To contrast this, let's look at how to make the worst queen. If you are doing emergency queens you have no real control over the larvae they pick since you aren't grafting them. If you are grafting, then picking a large larvae will result in an intercaste queen.

If you set things up so the cell builder (or the queenless half of a split) ends up with a low density of bees (a small nuc set in a new place so all the field bees leave or a hive that is weak to start with) then the queen will likely be poorly fed. Also if you do this in a dearth and don't feed she will be poorly fed.

If you do this at a time when there are few to no drones (too early, in a dearth when the bees have killed off the drones etc.) then she will be poorly bred.

If you put the cells in mating nucs and catch them as soon as they lay an egg, you will interrupt the development of their ovarioles and get a poor queen.

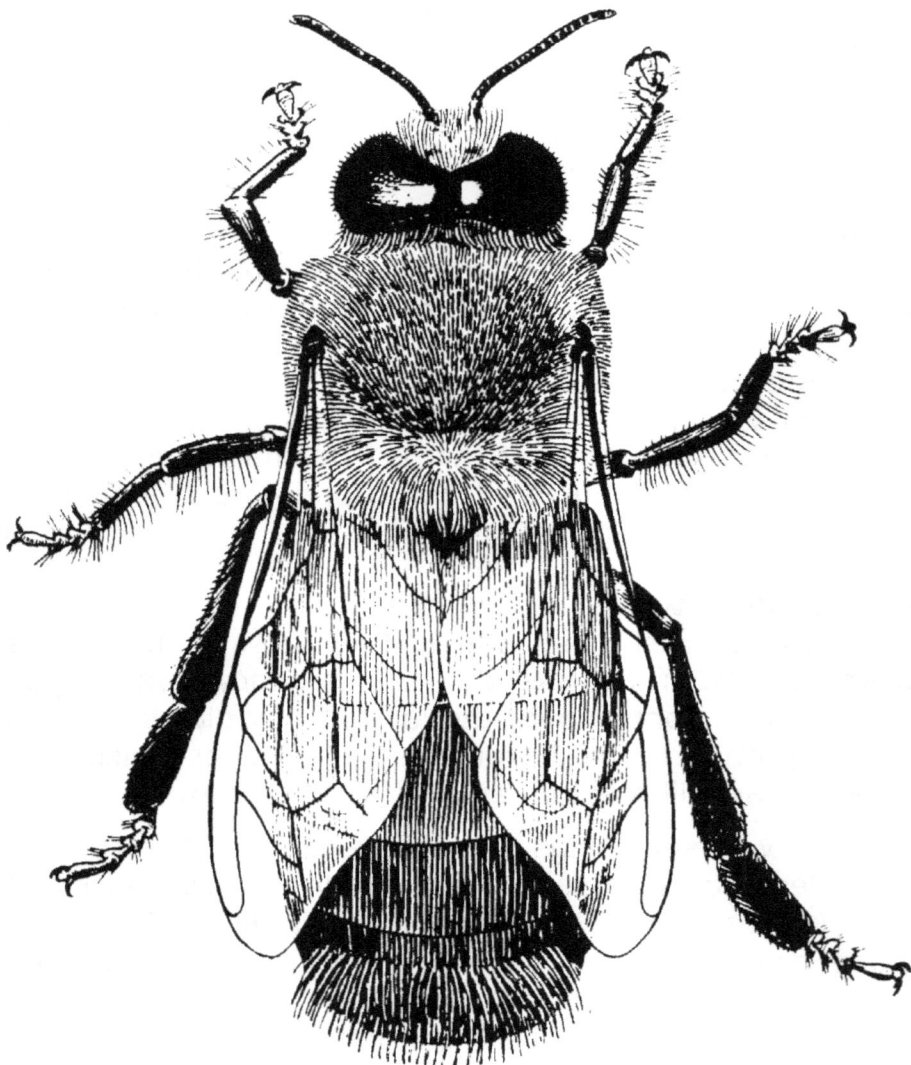

Drone (male) bee

Timing

Timing is essential with queen rearing. You want to raise queens when there are plenty of drones available. If you don't see drones flying, it's too early for three reasons. 1) you need mature drones to mate with the queen. If they are flying now, they will be mature enough when the queen is mating. 2) drones are an indicator of the willingness of the bees to raise queens. If they are not raising drones than they are not ready to raise queens. 3) the presence of drones indicates there is adequate forage that the bees can afford to spend to raise drones. If they can afford to spend it on drones they will be willing to spend it on queens. Also, you need to have enough bees as resources to make mating nucs and to populate starter hives (swarm boxes). If there are no drones yet, it's too early in the spring. How late you can continue to raise queens is hard to say. I have had the bees lose interest as early as July and as late as October. Once the bees lose interest it's just a waste of effort. While we are on the topic of drones flying, keep in mind that drones are essential to the process and lots of drones are essential to the quality of the queens. I use natural comb and my bees have as much drone comb as they want. I never cull drone comb. I just move it to the outside edges. Typically my hives have 20% drone comb and in the peak of swarm season they will have 20% drones in the colony. This is a good thing.

Eggs

Eggs close up

Section 2: Equipment

Basswood or Linden

Equipment you will need

Now that we have laid the groundwork for the concepts of queen rearing, let's look at what we need for equipment for my method of queen rearing. We will start with a list and then talk about each item.

- Swarm Box/Starter Hive
- Grafting tools
- Cell bars
- Cell cups
- Easel
- Lights (especially a bright flashlight)
- Mating nucs
- Marking pens
- Hairclip queen catcher
- Marking tube
- Queen muff

Lots of bees in the air

Swarm Box/Starter Hive

The swarm box was invented, and named by Henry Alley. It is a misnomer in the sense that it has nothing to do with swarms. So Jay Smith, a fan of the "swarm box" decided in his Better Queens book to rename it a "Starter Hive". The concept of a starter hive is that it's narrow (usually five frames wide) and tall (usually taller than the frames being used so you can put sponges for water on the bottom and so they get good ventilation), is well ventilated with screen wire around the bottom so you can crowd the bees into it without them suffocating, and preferably bee proof so you can put it in a cool basement without bees getting into your house. You can build your own, and I have in the past, but lately I just buy them from Betterbee.com

Betterbee swarm box

Jay Smith's Starter Hive from Better Queens

Jay Smith's Swarm Box from Queen Rearing Simplified

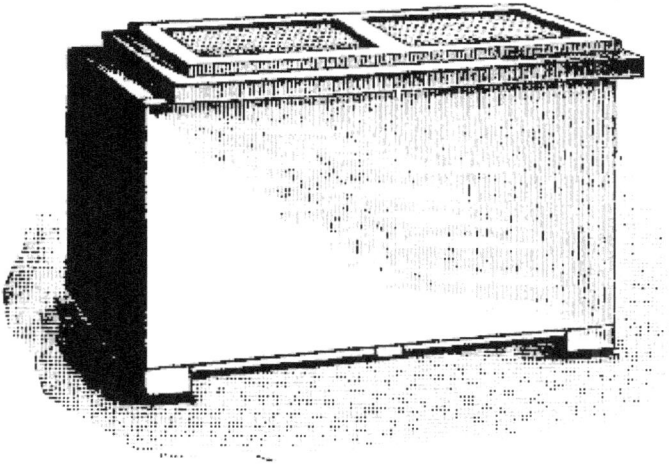

Swarm box from Henry Alley's book

Putting wet rags in the bottom of the swarm box

Putting Some frames of pollen and nectar in the swarm box

Shaking a swarm box

Grafting Tools

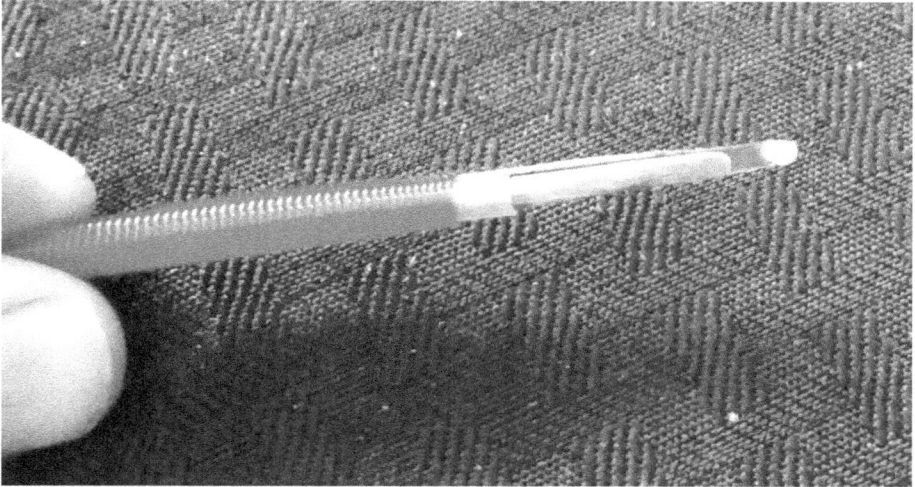

Chinese grafting tool with larvae

We could talk about a lot of grafting tools. I suppose the only real reason to talk about anything other than the Chinese grafting tools is the make shift ones that you could use without buying anything or in a pinch when you don't have one handy. So first let's talk about the Chinese grafting tool. The handle is usually plastic of some kind. The tongue that picks up the larvae and the royal jelly is made of horn. The plunger that pushes the larvae off is made of either plastic or bamboo. The quality control on these is not that great so buy several. After you get a feel for what makes a good one, you can pick through the ones at your local bee conference at the vendors tables, but until then just buy a dozen to start with and sort through them. You will also get better over time at tweaking them to make them work better. The things that are often wrong with them are:

- The tongue is too stiff. Some 600 grit sandpaper can be used to feather it out a bit, but don't get carried away because you don't want it too thin.

- The tongue is too thin and folds over instead of sliding under the larvae. Nothing you can do to fix this.

- The tongue isn't tight against the plunger and the larvae slides under the plunger. You can usually bend the tongue a bit to get a tight fit again.

- The plunger doesn't make it all the way to the end of the tongue or the tongue doesn't stick out far enough. Often you can get the tongue to slide a bit in or out to fix this.

The principle of this type of grafting tool is that the tongue slides under the larvae and royal jelly and picks up both. Then you lay it in the queen cell and bend it down a bit so the plunger is tight on the tongue and slide the plunger until the larvae is touching the cell. Then you pull the tongue out and the larvae is stuck to the cell.

Make shift grafting tools:

Here is a picture of a toothpick made into a grafting tool by Doolittle:

Other make shift tools have been made from solid copper wire, a stem of grass etc. but all of these have the same underlying problem which is that they only work with older larvae that you can get the tongue/blade to pick them up by the middle of the "C" shape. The Chinese grafting tool, on the other hand, will pick them up when they are too small to use other tools and bring the royal jelly along as well.

Cell Bars

There are many variations on cell bars. Most seem to be regular frames that have been converted by just adding bars to them. But I prefer this style which you can buy from Betterbee.com:

Except that I run all medium frames so I buy these from honeyrunapiaries.com:

The advantage to this style of cell bar frames is that they are only ¾″ thick (19mm) so I can fit two of them in the space that would otherwise be used by a standard frame with bars added. I buy some of these and I make some. In the past I have bought the deep ones and cut them down. Then I found the medium ones from honeyrunapiaries.com. The bars that hang in the frames have a groove in them for the cell cups. I use the JZBZ cell cups. You can make your own out of wax but this is a lot more work.

Cell Cups

The cell cup is something that, to the bees, resembles a queen cup. It is what you graft the larvae into for the bees to start queen cells. The simplest is to buy the JZBZ cups. Other options are to make your own wax cups. Usually these are done with a wood dowel dipped in wax. Then when they are done and pulled off of the dowel you wax them onto a wood disk and wax the wood disk onto the cell bar. That way you can pry the wood disk off without damaging the cell when you are ready to put them in a mating nuc. The JZBZ cups are available from most bee supply places. Color doesn't matter. They also sell plastic bars to put the JZBZ cups in. I find these pretty handy, but the wood ones that come with the frames work fine for a while until the plastic plug on the cups wears away too much wood. Then you need to put a bit of propolis or wax on the tip to get them to stick. I recommend the plastic cups but just in case you want to make your own cups here is Jay Smith's advice:

"Wax is saved from the year previous. For this a solar wax extractor is an important item. During the summer months, many small pieces of comb are found that can be thrown into it. This makes the finest cell building wax. In the nuclei, bits of comb are built and when introducing queens, where a frame is taken out, the bees will construct more or less comb. All these can go into the wax extractor. From the wax extractor, the wax is placed in small molds, for use in dipping queen-cells. I have enough cell bars to last the season, so we always dip sufficient each winter to supply us through the entire summer.

"Our cell-dipping outfit contains twenty cell-forming sticks, which work through holes made in two pieces of heavy tin. Metal is much better than wood since the latter swells when wet and the forming sticks do not work freely through the holes. These pieces of tin are fourteen or fifteen inches long, fastened one and one-quarter inches apart to small blocks of wood, which are to serve as handles when dipping the bars into the trays. Each piece of metal is pierced with twenty holes, one-fourth inch apart, and seven-sixteenths inch in diameter. The holes are exactly opposite each other on the two bars, in order that the cell forming sticks may slip up and down through them easily.

A solar wax extractor is an important item.

"Two trays are used, one five by sixteen inches, the other two and one-half by fifteen inches. Water is placed in the larger forming a double boiler; while wax is placed in the inner tray and the whole set over the heat. The wax should be kept at the lowest temperature at which it will remain liquid. If it becomes too cool the cells will be lumpy; if too hot, they do not slip from the sticks. If one is not experienced, it is well, when the wax apparently reaches the proper temperature for successful dipping, to try dipping one stick, and, if the wax proves of satisfactory temperature, proceed to work.

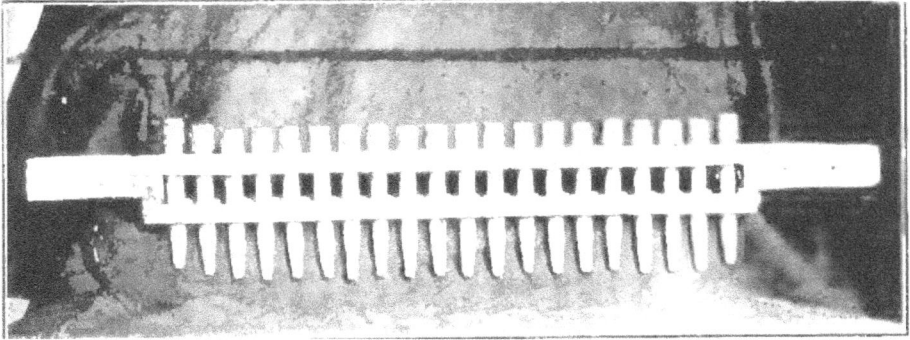

Our cell-dipping outfit contains twenty forming sticks.

Cells of the proper size and shape.

"First, dip the ends of the forming sticks in cold water, then dip into the melted wax; again dip in the water and back into the wax for about four dippings, care being taken to have a firm thick base, with a thin even edge. By dipping the sticks in the wax and holding the bar up until a drop forms on the base of the cell, a thick base is procured. A thick base is necessary, for in trimming off the cells with a knife the cells would be injured if too short. When completed, the cells should be about five-sixteenths of an inch across the mouth and one-half inch deep inside measurements.

"Many beekeepers make a mistake in believing that the most important feature for successful cell acceptance is the grafting of the larvae into the cells cups; but a far more important feature is that of making cells of the proper shape and size. The ideal cell would be as the bees build them, large inside, with a small mouth; but it is not possible, or at least practical for the beekeeper to make cells of this shape. Upon several occasions, I have given cells that had been accepted and slightly built out in the swarm box to a colony for finishing, when by accident it contained a virgin queen. Of course, the larvae and jelly were both quickly cleaned out. I have given one bar of such cells to a swarm box and two bars of our dipped cells. The bees seemed to concentrate all their efforts on the cells already worked on by the bees and neglected my dipped cells. The bees prefer to make the mouth of the cell just large enough for a worker bee to crawl into, and it is frequently noticed that sometimes in the workers haste to back out of a queen-cell when smoke is blown into the hive, it is caught and has to do considerable scrambling and kicking before it can get out. I find the best cell for practical purposes is one whose size is between that of the inside of a natural queen-cell at its largest place and the mouth of the cell,

this being five-sixteenths of an inch as given above. In our early experience, many of us, enthusiastic in rearing larger queens, sought to accomplish this by making larger cells; but being large at the mouth, the bees were loath to accept them, and it took considerable work on their part to build them over to the size they should be. When the bees get to work on the cells they mold them into the shape they want, regardless of the size and shape the beekeeper has made them. The smaller cells will give better acceptance than the larger ones; but do not for a moment imagine this cramps the larva and produces an inferior queen, for the bees enlarge the cell to suit their own fancy. For experimental purposes I have dipped queen-cells the size of a worker-cell, and excellent results were obtained. Cells larger than five-sixteenths of an inch are not accepted so readily as those of this size or smaller.

And the cell cups painted at the base.

"Nothing but pure beeswax of good quality should be used. Upon one occasion, when everything was going finely, cells accepted and built out nicely, the bees in the swarm boxes began to balk until accepted less than twenty-five per cent of those given. I had all conditions right, as I supposed, the same as before-plenty of young bees, well fed. At length I noticed the wax of which we made the cells was not so white as some we had been using. I made up a new batch of cells from clear white wax, and as if by magic, all cells were again accepted and everything went on splendidly as before. Instead of heating the wax in a double boiler as we do now, this wax had been set directly over the flame and had become slightly scorched and darkened, so the bees would have none of it.

"After the cells have remained in water long enough to become slightly hardened, they are loosened by giving each a slight twist, but allowed to remain on the sticks. They are then placed on the cell bar, the frame being supported on blocks. A small round paint brush is dipped in hot wax, and the cell cups painted at the base where they come in contact with the cell bar. A kettle should be kept at hand for melting additional wax to add to that in the inner tray, in order that sufficient wax may be had to make the cells the necessary one-half inch in depth. If the wax in ether becomes dark-colored or impure it should be discarded, and an entire batch of new clear wax placed in the tray. However, the darker wax may be used to paint the bases of the cells to cause them to adhere to the bar.

"When the wax has become thoroughly cool, the frame is lifted off and all of the forming sticks come out of the cells easily. If properly done, the cells will remain on the bars even if subjected to considerable rough usage.

When the cell bars are all finished they should be wrapped carefully in paper to be kept free from dust, since the bees will not accept dirty or dusty cells. If you have on hand the cardboard cartons in which foundation is shipped they make ideal containers for the cell bars.

Suggestions in Making Cell Cups.

"Of course it is not advisable for the beginner to have a dipping outfit made as previously described. After mastering the grafting method, he may enlarge upon his equipment as he wishes.

"The beginner can either dip his cells one at a time and mount them or he can purchase ready-pressed cells from dealers in bee supplies. Either one will give perfect results. These cells may be mounted on bars as needed, thus eliminating the necessity of purchasing a large number of bars. The base of these cells may be dipped in hot wax and stuck on to the bar when needed. To avoid the necessity of getting the swarm box, he can also use the queenless and broodless method described in Chapter XIII. However, I believe it pays to use the swarm box, for one can, as a rule, get better results. In this way it is possible to experiment until one gets his hand in without putting much money into equipment, and as he progresses can add to the

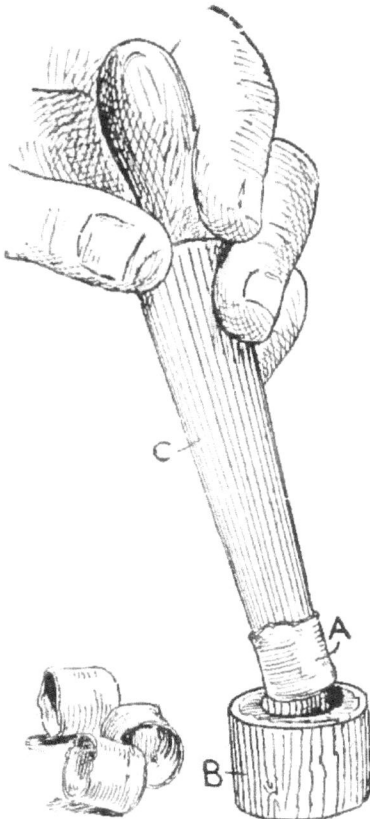

Pressed cell cup.

equipment to fit his requirements.

"If one has difficulty in making his cells, one at a time or collectively, he can use to advantage the ready-made pressed cells sold by all dealers. Where only a small number are required the beginner will probably do better to buy what few he uses. The making of dipped cells is a nice art, and unless they are made just right, the bees will reject them."

Easel

This is not necessary, but it is extremely helpful. You can buy small easels that are made either for painting or for holding pictures. You want something small and simple. The purpose of this is to hold the frame at the correct angle so you can see into the bottom of the cells. I buy mine usually on Amazon, but you can often find them at a hobby place.

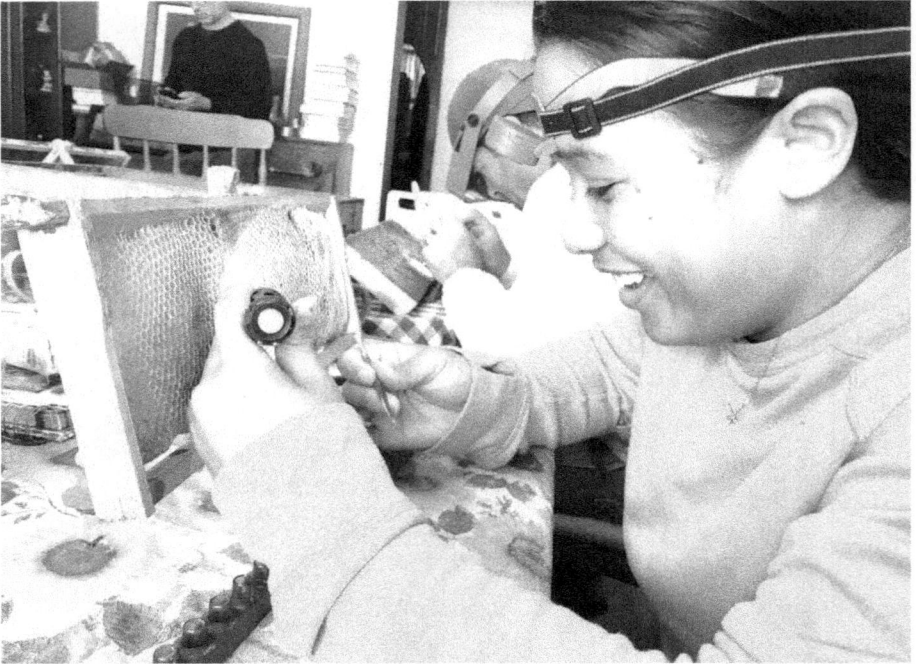

Grafting using the easel

Mating Nucs

Two By Four Mating Nucs

Splitting a ten frame box into four nucs with two frames each. Note the blue cloth sticking out. These are canvas inner covers so I can open one nuc at a time without them boiling over into the next nuc. Also note the Ready Date Nuc Calendars on the end.

A note on mating nucs

In my opinion it makes the most sense to use standard frames for your mating nucs. Here are a few beekeepers who agree with that:

"Some queen-breeders use a very small hive with much smaller frames than their common ones for keeping their queens in till mated, but for several reasons I consider it best to have but the one frame in both the queen-rearing and the ordinary hives. In the first place, a nucleus colony can be formed in a few minutes from any hive by simply transferring two or three frames and the adhering bees from it to the nucleus hive. Then again, a nucleus colony can be built up at any time or united with another where the frames are all alike, with very little trouble. And lastly, we have only the one sized frames to make. I have always used a nucleus hive such as I have described, and would not care to use any other."—Isaac Hopkins, The Australasian Bee Manual

"for the honey-producer there seems no great advantage in baby nuclei. He generally needs to make some increase, and it is more convenient for him to use 2 or 3-frame nuclei for queen-rearing, and then build them up into full colonies...I use a full hive for each nucleus, merely putting 3 or 4 frames in one side of the hive, with a dummy beside them. To be sure, it takes more bees than to have three nuclei in one hive, but it is a good bit

*more convenient to build up into a full colony a nu-
cleus that has the whole hive to itself."—C.C. Miller, Fifty Years Among the Bees*

*"The small Baby Nucleus hive had a run for a while
but is now generally considered a mere passing
fad. It is so small that the bees are put into an unnatural condition, and they therefore perform in an
unnatural manner...I strongly advise a nucleus hive
that will take the regular brood-frame that is used
in your hives. The one that I use is a twin hive,
each compartment large enough to hold two jumbo
frames and a division-board."—Smith, Queen Rearing Simplified*

*"I was convinced that the best nucleus that I could
possibly have, was one or two frames in an ordinary hive. In this way all work done by the nucleus
was readily available for the use of any colony, after I was through with the nucleus...take a frame
of brood and one of honey, together with all of the
adhering bees, being careful not to get the old
Queen, and put the frames into a hive where you
wish the nucleus to stand...drawing up the division-board so as to adjust the hive to the size of
the colony."—G. M. Doolittle, Scientific Queen-
Rearing*

Stack of mating nucs

Mating nuc with no frames

Calendar

You will need some way to track when things are due and the current status of your mating nucs. This becomes very important when you do one batch after another. I used to use Ready Date Nuc Calendar and you may be able to find something similar still for sale, though They stopped making the Ready Date version. I had a brand made up and I use it on my nucs as well as my hives. It looks like this:

1	2	3	4	5	6	7
8	9	10	11	12	13	14
15	16	17	18	19	20	21
22	23	24	25	26	27	28
29	30	31	Egg	Opn	Cpd	Fsg
Mkd	Gnt	Def	XQ	VQ	LQ	Gd
Swm	Spr	Eng	Dx	↑	↓	Flp

The numbers are for dates. I laid them out in weeks so you can count weeks by just moving down the column. I usually use the same date for the batch all the way through and that is the date I expect to have a laying queen. The rest of the items are the current state of the mating nuc or colony. After each inspection

I can mark if I saw eggs, open brood, capped brood, festooning bees, a marked queen, a gentle colony, a defensive colony, queenless, virgin queen, laying queen, grafted cell, swarm cell, supersedure cell, emergency cell, disease, increasing, decreasing, and the last is to flag it for a followup, which means that something needs to be checked or done. Maybe it needs a super or it should be fed or it needs more bees in the nuc. Things are marked with push pins. Here we are branding boxes:

Here is Doolittle's design using pins to mark things. You take two long pins and bend the tips and put them into the hole in the center of the circles. If it has a bit of a sharp head on it, like a straight pin, then you should be able to push the head in enough to hold it in place.

QUEEN REGISTER.

EGGS.

No. . . . . . .

MISSING.   **BROOD.**

LAYING.   o   **CELL.**

APPROVED.   **HATCHED.**

NOT APPROVED.

DIRECTIONS.—Tack the Card on a conspicuous part of the Hive or Nucleus box ; then, with a pair of Pliers, force a common pin into the centre O of each circle, after it is bent in such a manner that the head will press securely on any figure or word.

MARCH.

OCT.   **APRIL.**

SEPT.   O   **MAY.**

AUG.   **JUNE.**

JULY.

Of course another choice is to just keep a calendar in the house and use a color coded pin in the box to tie it to a particular batch which is tracked on the calendar in the house.

bushfarms.com

1	2	3	4	5	6	7
8	9	10	11	12	13	14
15	16	17	18	19	20	21
22	23	24	25	26	27	28
39	30	31	Egg	Opn	Cpd	Fsg
Mkd	Gnt	Def	XQ	VQ	LQ	Gft
Swm	Spr	Emg	Dx	↑	↓	Flp

Calendar in use

Brush

Because queen cells are fragile, shaking bees from any frame that contains queen cells, especially a cell bar, is not recommended. A brush is preferable. A large feather works fine for this as do the commercially available bee brushes. In a pinch a bunch of grass can work.

locust blooming

Hairclip Queen Catcher

It is certainly worth practicing catching queens with your bare hands by practicing on drones. But it is still worth having a hairclip queen catcher. Not only does it help when catching queens, but it is a way to keep track of the queen. Once you have her caught, be sure you keep the clip on a box of her bees so other bees don't attack her. I have made the mistake of putting one on top of the hive next door and lost the queen. These are available from the various queen supply houses. I don't care for the metal ones. The one pictured is kind of soft plastic. If you are gentle when closing it, it works well. If you get too careless you can cut the queen in the hinge side like scissors or smash her when you first push it down on the comb. You need to learn to be fast and gentle, just like when you catch them barehanded.

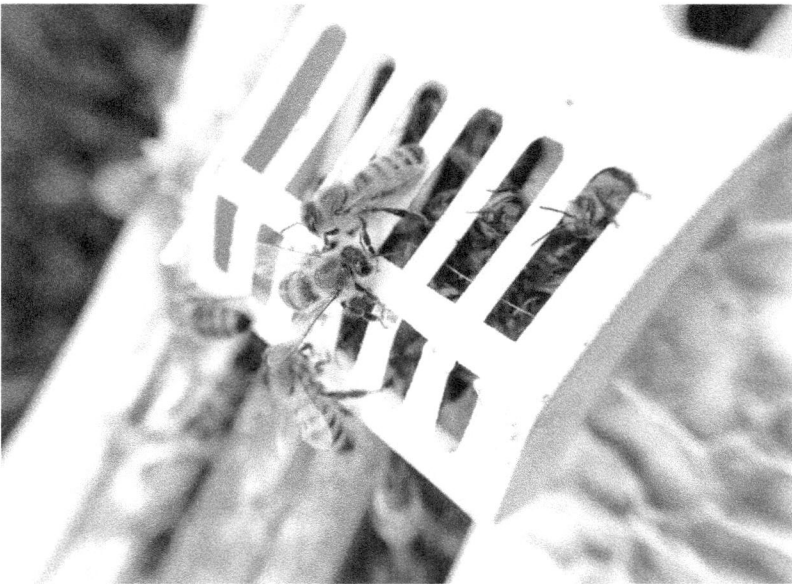

Hairclip Queen Catcher

Marking drones for practice with next year's color

Hair clip queen catcher

Marking Pens

My preferred marking pens are the Testors enamel pens from the hardware store. The beekeeping supply places sell various kinds. As to colors I prefer the brightest I can get. A florescent color is much easier to spot than a regular color. You can usually find florescent yellow, red, green and blue. White, is of course, white. It's a good idea at a minimum to have this year's color and next year's color. Next year's color is for practice on drones. By next year this year's drones will all be dead and you probably don't have a lot of five year old queens. This year's color is for the queens you are raising now. You might find it useful, if your queens aren't currently marked, to have last year's color to mark your existing overwintered queens.

Queen marking colors:

Years Ending in:

- 1 or 6 – White
- 2 or 7 – Yellow
- 3 or 8 – Red
- 4 or 9 – Green
- 5 or 0 – Blue

Marking drones for practice with next year's color

Marking Tube

Queen marking tube

One way to control holding a queen while you mark her is a tube like this. You use the plunger to pin her to the screen and mark her through the screen.

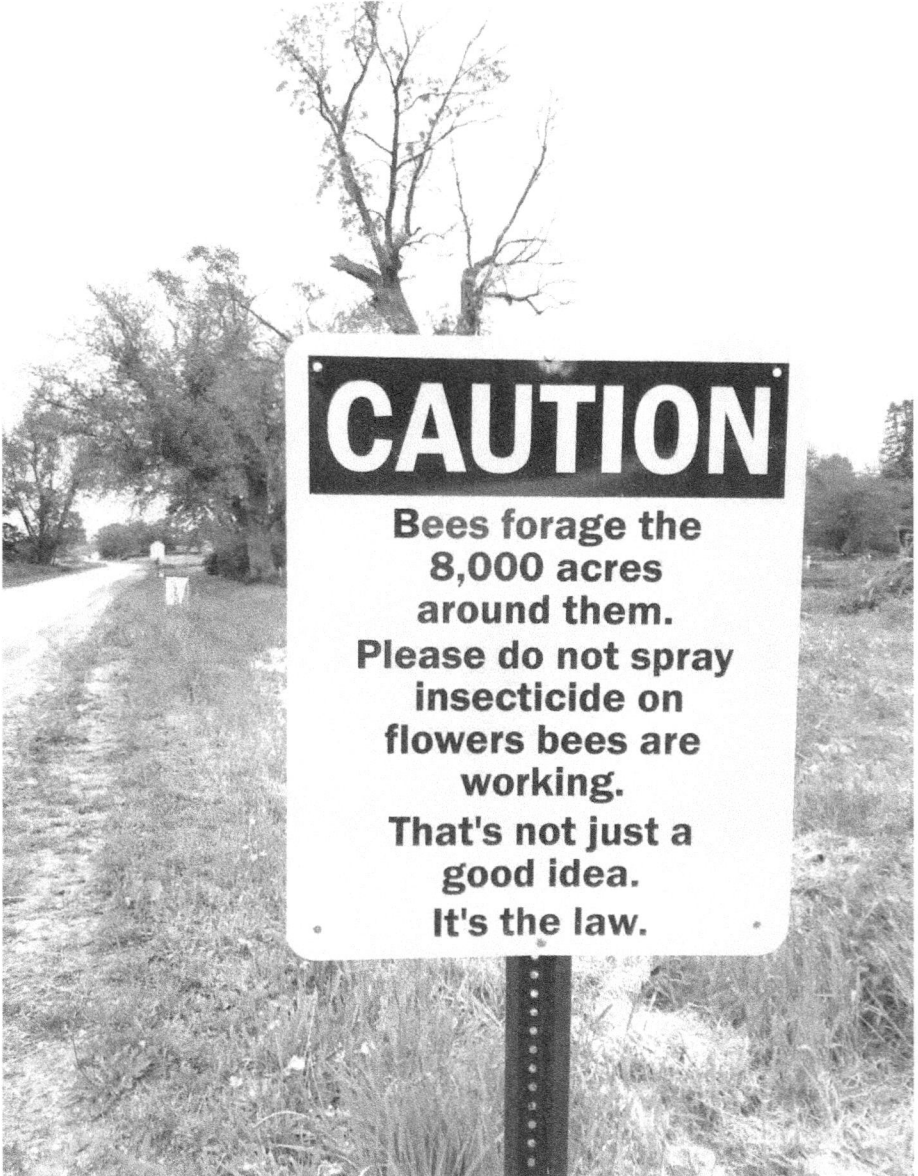

CAUTION

Bees forage the
8,000 acres
around them.
Please do not spray
insecticide on
flowers bees are
working.
That's not just a
good idea.
It's the law.

educating the neighbors

Queen Muff

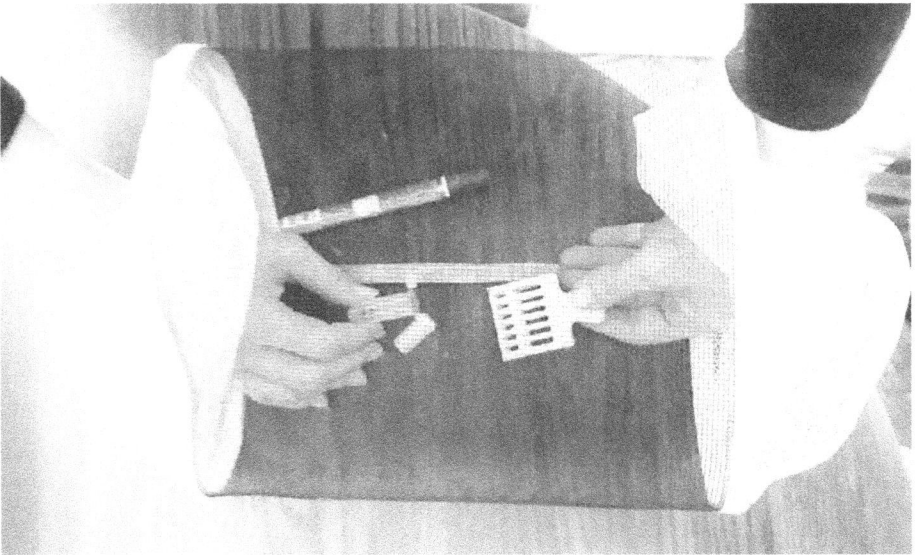

queen muff

This is a device used to keep the queen from flying off while you are marking her or otherwise manipulating a queen. It's not necessary, but it can be quite useful. Your hands go into the elastic cuffs and you can see what you are doing through the screen.

Huber Leaf Hive

Section 3: Method

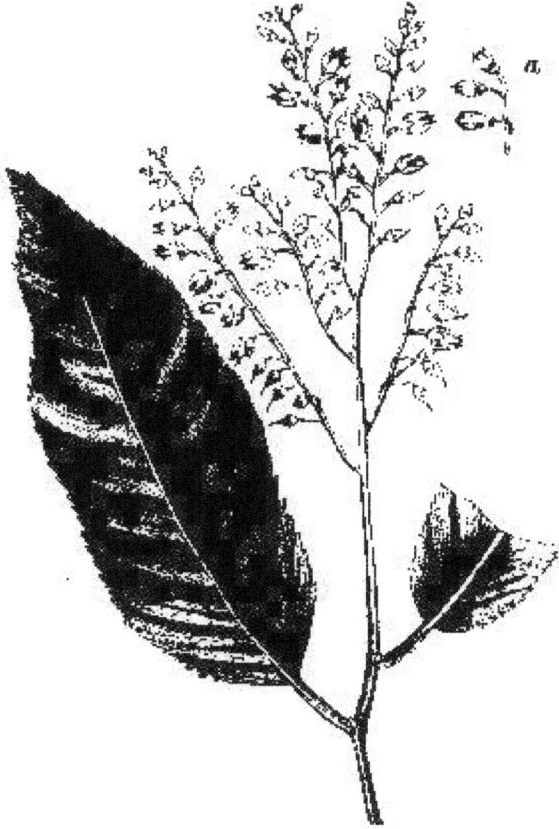

Sourwood

Overview

The basic steps of queen rearing are:

- Setup the cell starter

- Find the right age larvae from the preferred queen to graft from

- Put the larvae in the starter to get them flooded with royal jelly

- Graft

- Put the grafts in the cell starter

- Setup a cell finisher

- 24 to 48 hours later, move the grafts to the cell finisher

- 8-10 days after grafting, set up mating nucs and put the cells in the mating nucs

- Two weeks after putting the cells in the mating nucs, check to see if you have laying queens.

This is slightly different when you are doing subsequent batches of queens. I typically do two batches at the same time, mostly as insurance in case something goes wrong with one of them, and I do this every week. So once there are mating nucs set up and there are queens in them, then I'm not so much setting up mating nucs as managing them and catching queens to make room for new queen cells. So the "set up mating nucs" step becomes, catch queens, boost weak mating nucs, split or move strong mating nucs so they don't swarm and then put cells in available mating nucs.

Starter Hive

Setup the Cell Starter

As we previously mentioned the way we are doing our starter is with a "starter hive" or "swarm box". The process for setting up a "swarm box" is first to put two or three large sponges soaked in water on the bottom. Then start going through strong hives that we can shake bees from. While we are going through these hives we are looking for any of the following:

- Queen (so we can set her aside and not get her in the swarm box)
- Frames with no brood but nectar
- Frames with no brood but bee bread
- Brood frames with no queen on them

As we find frames with bee bread or nectar (or a mixture of both) we will put these in the swarm box with the adhering bees. Before I start shaking bees I would like to have at least one frame in the box, and preferably two. These I will put on the outside edges so I can shake bees into the middle. The swarm box I bought from Betterbee has a cross piece inside a ways down to slam frames on to knock off the bees down inside the box. So once I have the outside frames in (for the bees to hang from and climb on) I start shaking in bees from brood frames. I don't want to weaken any hive too much, so I will spread out how many bees I shake from any given hive. My goal is to end up with sponges soaked in water on the bottom, bees spilling over the sides, two frames of bee bread and two frames of nectar. The most common mistake is to not put enough bees in the starter. It should be pretty much full of bees. Every frame should be densely covered and a lot of festooning bees hanging from the bottom bars of the frames.   The object of this swarm box is twofold. First we are shaking in nurse bees (they came from brood combs) that were just feeding larvae so they were producing bee "milk".   Second, there is no brood in the box to feed, so not only are they queen-less, but they have an abundance of royal jelly (bee "milk") and nowhere to put it.   That way when I give them queen cells they will jump on them and feed them well. So the benefit is that I will get a lot of cells start-ed and those cells will be really well fed.

Put this swarm box in a shady place.   Remember that the sun moves around so a place that is shady all day long is difficult to find. If you have confidence that the box is bee proof, you can put it in a cool dark basement.

Putting starter hive in the shade

Find the Larvae to Graft From

If one of the hives I'm getting bees for the starter hive from is the one I want to graft from I can keep an eye out for the right age larvae. If not, then I will open the donor hive and look for the right age larvae. A good experiment to learn how old larvae is would be to build the #5 hardware cloth push-in-cage pictured in the chapter titled "Queen Rearing Concepts" and use that to confine the queen on some drawn comb and put this in the middle of the brood nest. Be careful to leave at least 3/8" space between the top of the cage and the next frame so bees can get between. If you leave it 24 hours and then look you should see eggs. If for some reason the queen has not laid eggs come back in 24 hours and check again. Once you have eggs, you know that they are 24 hours or less in age. If you come back 3 days after that some of them will likely be hatched and some will not. The just hatched ones will be a tiny drop of royal jelly with an imperfection on the surface that is too small to tell what it is. If you come back again in 12 hours you will see these are larger. If you check back in another 12 hours you'll see what 24 hour old larvae look like. Anything bigger than this is not going to make as good of a queen. In another 12 hours you have 36 hour old larvae and this is the oldest you should *ever* graft. I try to get them not more than 12 hours old. If you don't want to do this learning experiment, then just look for eggs and look at the edge of where the eggs stop and some of them are a drop of shiny liquid in the bottom. These are just hatched eggs. This is the age you want. Typically the frames you are looking for are partly eggs and partly the right age larvae and possibly some too old larvae. Ideally I like to find two frames of the right age and put these in the Starter Hive for an hour or two. That way they will be flooded with royal jelly and that makes them easier to graft with the Chinese grafting tool and they will be better fed. If they look pretty flooded already, I might skip this step. After they have been flooded with royal jelly (either because they already were or because I put them in the starter) I brush all the bees off of them and take them to the house to graft. My house happens to be just across the yard from the bees, so this is the simplest for me. But when I have grafted in an outyard I have been know to do it sitting in the front seat of my van. It is easiest though if you can control the wind, the temperature and the light which is easier to do in the house. If you really want to get into queen rearing it might be worth building a little portable building to graft in with a big window for light and small enough you can keep the humidity up by sprinkling water around the building.

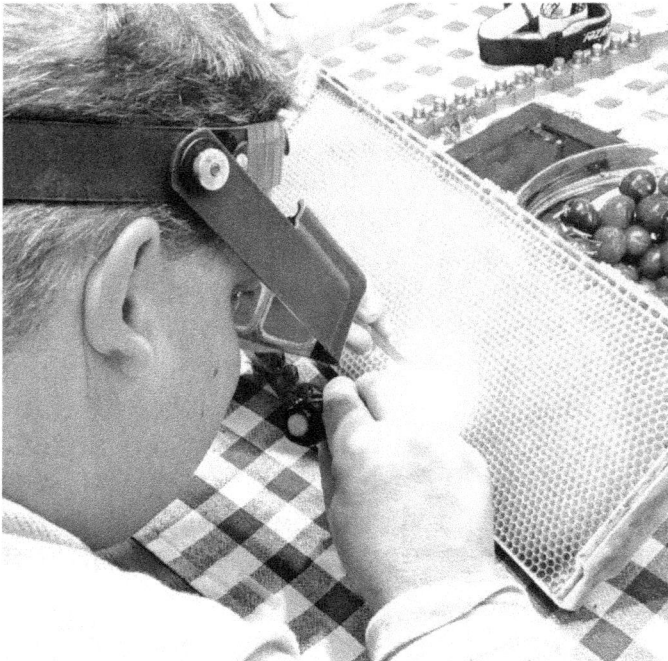

Grafting

Setting up

Now that we have the frame or frames of larvae in the grafting location (my dining room table in my case) it's probably summer and I probably have some fans around. These I need to turn off. The biggest danger to the larvae at this point is that they get dried out so the fan is a problem. Some wet cloths should be put on whatever frames of larvae you're not working with to keep them moist. I find that the angle of the cells seems to work out better if I turn the frame upside down on the easel. Then I adjust the angle (the leg of the easel) until I can easily see down into the cell. I like to use a flashlight to shine down into the cell, but if you have a reading lamp or something similar that you can get adjusted so that light goes down into the cell that can work. My dining room table is in a sort of box window so there is light coming in from three sides if I set up at the end of the table. This also helps.

Picking a grafting tool

Look through the grafting tools and pick one that has a nicely feathered tip that is flexible and a plunger that stays in contact with the tongue. If you have trouble grafting with one tool, change to another. When you find a good one, use a queen marking pen or a magic marker and mark that one so you can find it again.

Picking up the larva

It takes practice to get good at this. Don't get discouraged if it seems difficult. Once you get the hang of it, it's quite easy, but until then you'll have to think about what you are doing. You want the tongue of the grafting tool to slide down the outside of the cell. The "upper" side of the tongue is the side that the plunger is on. The "upper" side should face the middle of the cell. Find an angle that is comfortable and let the tongue slide down the side to the bottom where the tongue should bend and slide under the larvae. If it's old brood comb or plastic there should be no issues with the tongue poking through the cell, but on new comb this can be a problem. If you can, of course, pick old comb for this. If the tool is properly flexible the tongue should round the corner and slide under the larvae. Now the next tricky part is to lift it out of the cell without the larva getting stuck to the side. This requires a slightly angles movement where you move the tool back and up at the same time. If you fail at this and the larva touches the side it will almost always stick to the side and any attempts to get it out after that are liable to damage the larvae. Also you may get the larvae upside down. The idea is to slide under it and pick it up and set it down and push it just enough that it touches the bottom of the cell and it sticks there and you pull the tongue out from under it. This insures that the larva is right side up. If it is upside down it cannot breath. So be sure not to flip it over. Try to put it in the exact center of the cell. If you don't get it in the exact center don't worry about it, but ideally it should be in the middle. I like to set up about four bars (they come out of the frame) and lay a grafting tool at the point where I'm working to keep my place. I graft four and move the grafting tool over. When I'm done with all four bars I put them in a frame, cover them in a wet rag and do four more.

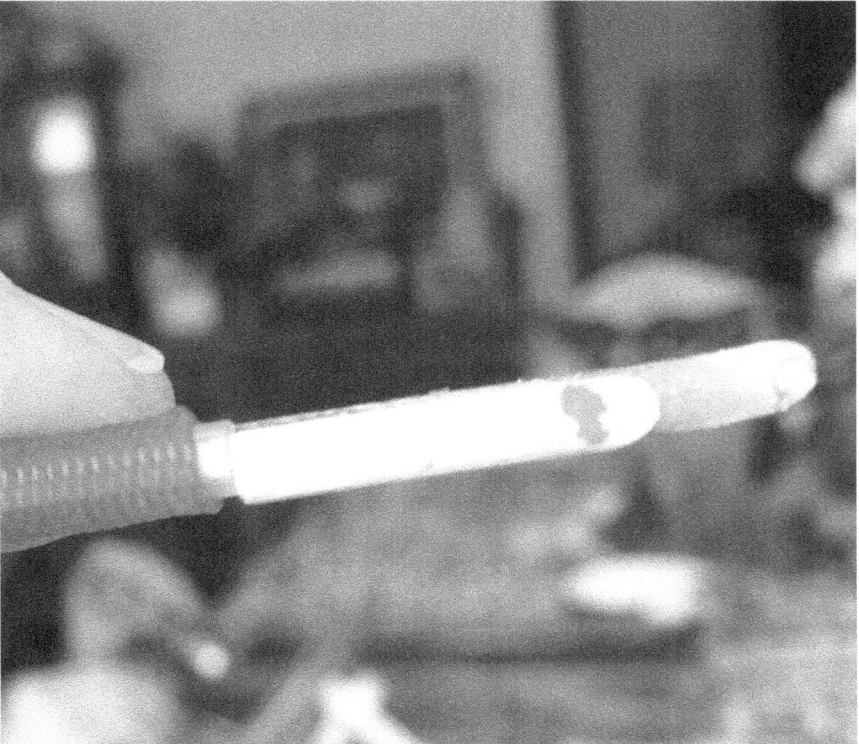

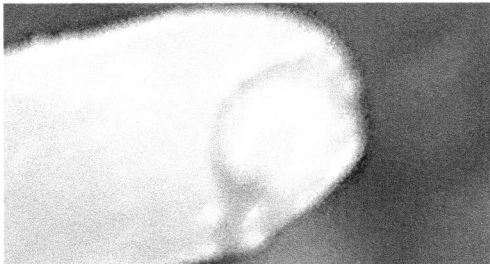

The right age larvae

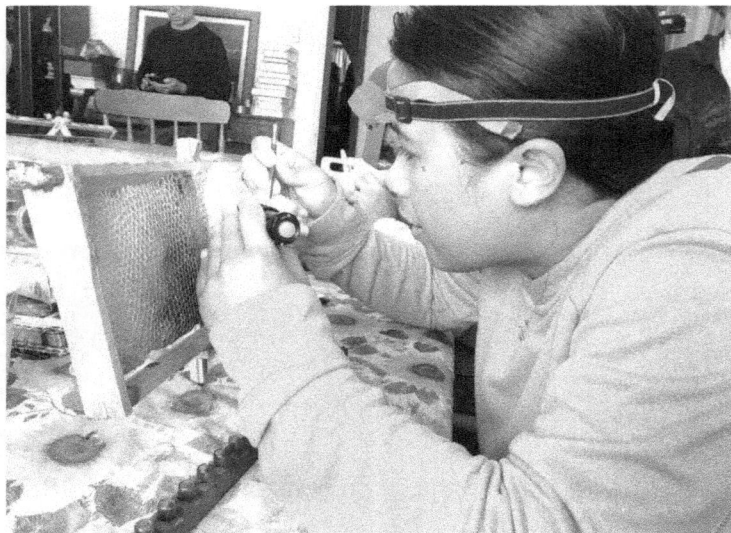

Grafting

Put the Cells in the Starter

Now that we have the larvae in the cells we need to put the cell bar frames with the cells into the starter hive (swarm box). We also need to return the frames we grafted from back into a hive somewhere. Probably the one we got them from but we could also give them to some other hive if we want. Any hive that is suspected of being queenless would be a good candidate to give these to. Depending on the design of your "swarm box" you may need to knock the bees off of the lid before you open it up as bees will be clinging to the top and festooning everywhere. The one I have from betterbee.com has a slide in lid that they fall off of when I open it without me having to shake them down. The two frames of cells go in the center of the swarm box. This is our starter. I need to keep it cool and in the shade for the next 24 hours minimum. If the weather is cool I leave them for 48 hours. If it's hot I don't go over 24 hours.

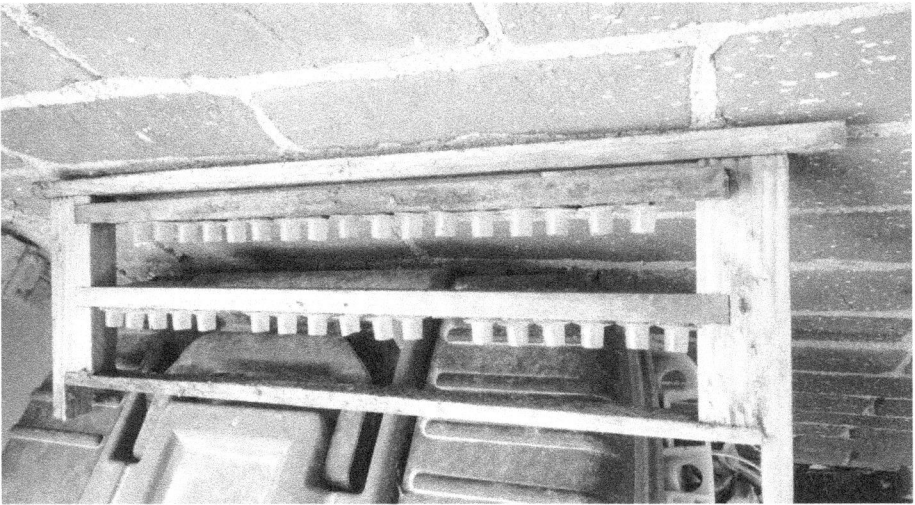

Cells

Grafting

Setting up a Cell Finisher

Pick a strong colony. It doesn't need to be your strongest (which will probably make you a honey crop) but it shouldn't be a weak colony. If I'm in a hurry all I really need to do is make sure the queen is down below an excluder at least one full box away from where I'm going to put the cells, but ideally I like to sort the entire colony.

Sorting the colony

I like to sort everything and then arrange it the way I want to. So I set several bottom boards on the ground and put empty boxes on them. I sort everything into the following groups by looking at each frame carefully and also looking for the queen:

- empty or mostly empty
- mostly capped brood
- mostly open brood
- mostly honey or bee bread

Now that I have them sorted I stack them back up on the original stand. I put the honey and bee bread on the bottom just so I won't have to lift it every time. Then I put the empty comb with one frame of open brood and the queen on next. Then a queen excluder. Then a box of capped brood. Then the open brood. Things don't ever come out even so some frames that don't entirely fit the description may end up somewhere different to fill out the box or to put the leftover frame somewhere. What I need to end up with is space for the cell bars in the top box which I will temporarily fill with empty frames. So now the hive has the queen down below with food to feed the larvae and space for her to lay (the empty combs). Then we have an ex-cluder to keep her below. Then we have the box of capped brood to make some space between the queen and the cells. Then we have the open brood where we will put the cells and the open brood will bring the nurse bees that I need to feed the queen cells. Once a starter is set up this way, since it is a queenright finisher it won't get burned out raising queen cells like a queen-less finisher does. The next round of cells will be a week from now and I can rearrange things by the box. The box the queen was in should be full of open brood so we will put it on top. The box of open brood which was on the top should now be capped brood, so that will go just above the excluder. The box of capped brood should now be emerged and mostly empty, so that will be the box below the excluder which we will need to put the queen in again. In this way the hive gets age grad-ed over time and we can keep it strong by keeping room for the queen to lay and save sorting repeatedly because the excluder and our timing will do that work for us. Since I usually do two batches at a time, of course, I will set up two finishers every week. Again in the ongoing cycle, once I have finishers set up, then next week I rearrange the boxes and put last week's cells in the mating nucs before putting this week's cells in the finisher.

Putting the cells in the finisher

Now that we have a finisher set up probably the next day or the day after (depending on the heat) we need to put the cells that are in the starter hive (swarm box) into the finisher. If we don't have room then we need to remove some frames. Hopefully we had it set up with two empty frames and we will pull those out and replace them with the cell bars. These will stay in the finisher until at least a week after they were graft- ed. Usually eight to ten days depending on the weather and my time.

The bees in the starter can be put in either the finisher or used to boost any mating nucs that are short on bees.

Finished cells

Finished cells

Setting up Mating Nucs

In theory my plan is to put the cells in mating nucs ten days after grafting. Since I only have weekends to work on bees usually, it usually comes out to more like eight days after grafting. In hot weather queens sometimes emerge 10 days after grafting, so eight days doesn't work so badly. But ideally in moderate weather ten days would be the goal and in hot weather nine or even eight would be safer.

We talked about the size of the nucs and I recommended using the size your brood frames are. Since I run all mediums it's also the size my honey supers are. So for the first round of mating nucs I'm looking for a frame of brood with bees (and no queen) and a frame of honey and a shaken brood frame of bees (just the bees not the frame). The shake of bees is to make up for drift. This is somewhat variable as on a hot day there are very few bees on the brood combs and I might even shake three frames of them into the nuc. On a cool day the brood frames are often thickly covered in bees and I might not shake any bees in. After a while you get a feel for the right amount of bees in the nuc. But for a rule of thumb, think in terms of twice as many bees as you think they need because half of them will fly back to the original hive. It's probably a good idea to count cells in the finisher just before you set these up so you don't have too many or too few.

Putting cells in the mating nucs

I used to set up the nucs the day before I put the cells in, but mostly now I just put them in when I set them up. Partly this is just a more efficient use of my time, but also I find that a mating nuc with a queen cell doesn't tend to get robbed as often as a mating nuc that is queenless.

The term "nuc" is short for "nucleus" as in "nucleus hive". The concept of a mating nuc is that the queen will emerge into this small amount of bees and become their queen. She will harden, fly out and mate, and start laying eggs while in this mating nuc. If we left the cells in the finisher, the first one out would kill the rest.

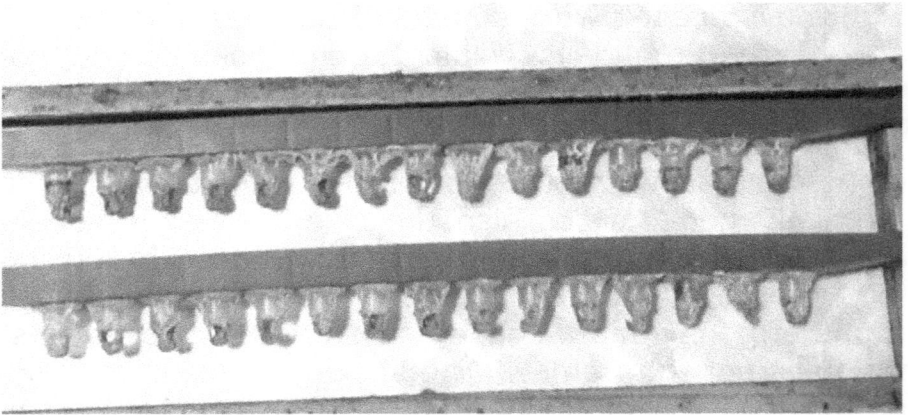

Cells killed by an early emerging queen

Checking the Mating Nucs

Since my schedule generally runs with a new batch of queens starting each week and since a queen cell that never emerged would give me another mating nuc I can put a cell in and since being queenless to long is hard on mating nucs, I check them a week later just for cells that have or have not emerged. If the cell did not emerge from the last batch, I put another cell in from the current batch. I don't look for the queen and I don't look for eggs (it's too early). The next week (two weeks after I put the cell in) I look for eggs and if I don't see eggs, I look for the queen. If I find the queen but no eggs, I let them go another week. If I find the queen and eggs, I mark it on my calendar and I wait another week to let her ovarioles finish developing. After they have been laying for at least part of two weeks, I will catch them and ship them out or bank them. Or, if I don't need the mating nucs yet, I might let them lay another week.

Checking mating nucs

Section 4: Management

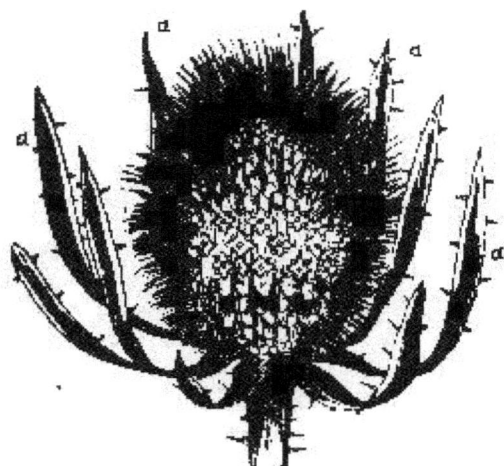

Teasel

Weekly Management

Now that we've gone through the details of each step, let's see how they fit together into a weekly schedule. Basically I start a new batch every Friday. In the summer the days are long enough I could do this after work on Friday, but lately I've been working four day weeks and have Friday's off. So here is my weekly management starting with Friday.

Friday

Setup a swarm box

Graft

Put the grafts in the swarm box

Saturday

Setup a cell starter if we need one

Rearrange cell starters if they are already going

Go through the mating nucs looking for un-emerged queens or queenless mating nucs that need cells. Also look for queenright mating nucs that are getting too strong and might swarm if left alone.

Catch queens and get them ready to ship or bank them.

Setup any extra mating nucs we will need for the number of cells we have

If the weather is hot, move the cells from the starter to the finisher.

Sunday

If the weather was cool, move the cells from the starter to the finisher.

Boost any mating nucs that are short on bees with the bees from the starter.

Put cells from last week in mating nucs.

Swarm

Managing Mating Nucs

When going through the mating nucs every week you need to look for these things:

- Are they strong? If they are strong they probably do have a queen and we won't need to add bees. If they are too strong they will swarm soon so I try to use that mating nuc to boost a weak one. If they are not strong then we need to boost them. This can be done in a variety of ways depending on the currently available resources. Often I move a frame with the queen on it from a strong mating nuc to one that has no queen and move a frame from the weak mating nuc to the strong one that used to have a queen (that I just took). All of the field bees from the strong nuc will now be flying back to the same place but with no queen there anymore they can be given a cell to get mated. Meanwhile the queen and the brood that went with her, in the weak location will quickly build up again. Another method is to shake some of the bees from the starter hive into the nuc when moving the cells to the finisher.

- Is there a virgin queen there? Sometimes they could be somewhat weak and still have a virgin queen. I don't try to catch those because they often fly if I try to. Virgins tend to be pretty small and flighty. They like to hide too. Just because I can't find one doesn't mean there isn't one. If they are somewhat calm and not too weak, I usually assume there is probably a virgin. But I use the calendar to keep track of when I was expecting a laying queen (two weeks after I put them in the mating nuc) and if I don't see a laying queen in two weeks I will give them a new cell.

- Is there a laying queen? Eggs would be a good enough indicator. If there are just some eggs and no open brood and no capped brood or only a little that I probably put there when I put a frame of brood in the mating nuc to set it up, then I need to wait a couple of weeks before catching the queen. I need her ovarioles to finish developing before I cage her.

- If I see the laying queen does she look flighty? Some queens sort of flap their wings as they run across the comb. I quickly put these back in and leave them. If they are acting flighty they will probably fly if I try to catch them. I'll come back next week and see if they have calmed down. If so I will mark them then. Also flighty queens are probably not fully developed yet.

- Laying workers often become an issue. With such a small colony, usually swapping all the frames from the laying worker queenless one with a queenright mating nuc and adding a cell will straighten things out. One skill you often learn if you pay attention is spotting the early stages of laying workers. Before you start to see multiple eggs you often see spotty brood with the only caps being drone cells and sometimes even some queen cells in a very weak mating nuc.

Bottom line

If you are pretty sure they have a queen (they still have a lot of bees and are behaving like a colony with a queen) and they have open and capped brood from this queen, pull the queen, ship her and put in a cell; if all those things are true except there aren't a lot of brood yet, wait another week. If you are pretty sure they are queenless (dwindling population, lethargic colony, grouchy colony) add a cell.

Shipping Queens

I assume you are selling queens if you bought this book. Perhaps you're only providing queens for yourself, but if you are selling them you will need to ship them.

Legalities

One of the first issues is legalities. Most states only allow you to ship queens if you have a health certificate issued by your state. This will vary from state to state, but if you have an apiary inspector in the state they should be able to tell you what you need to do. If you can't find them, try the department of agriculture for your state. My state doesn't have a regular inspector. A beekeeper who is in the entomology department does the inspections and I have to pay for them. Some states actually require you to buy a permit from that state to ship queens to that state. If this is the case it's not worth selling queens there at all. I would just not sell them. Unless you have built up such a reputation that you are selling hundreds of queens in that state it is not worth the cost let alone the effort. But most states you just need the certificate from your state.

Cages

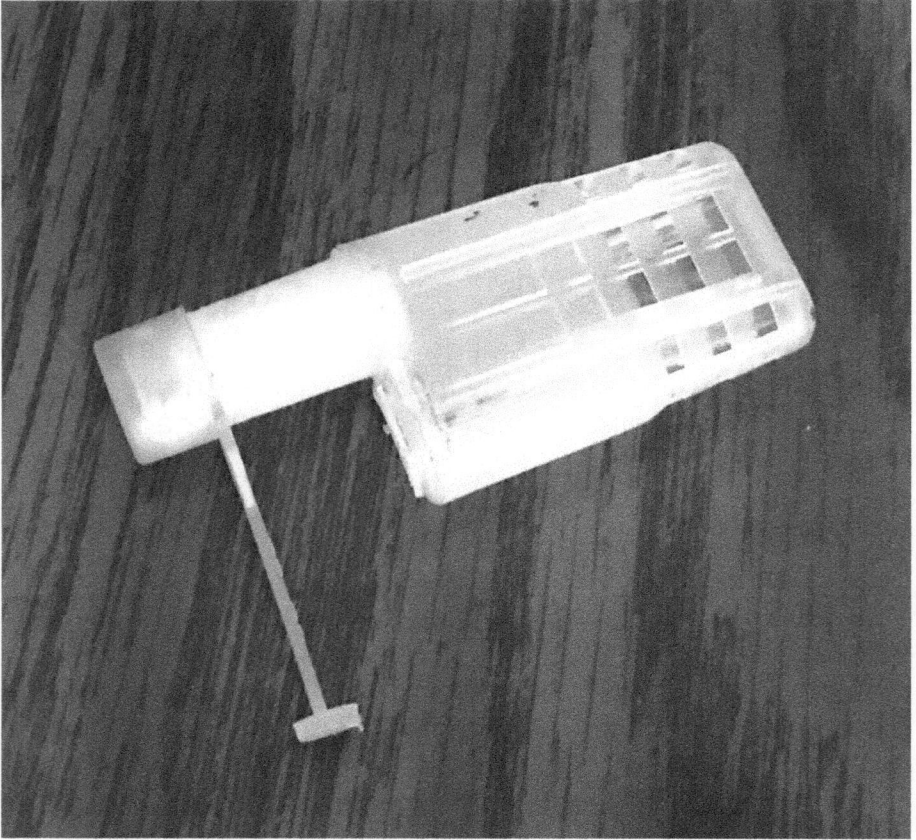

JZBZ cage

This book is about how *I* do my queen rearing and breeding and shipping. The only cages I use are the JZBZ cages. There are other types of cages but I find the JZBZ the simplest for me to use.

Candy

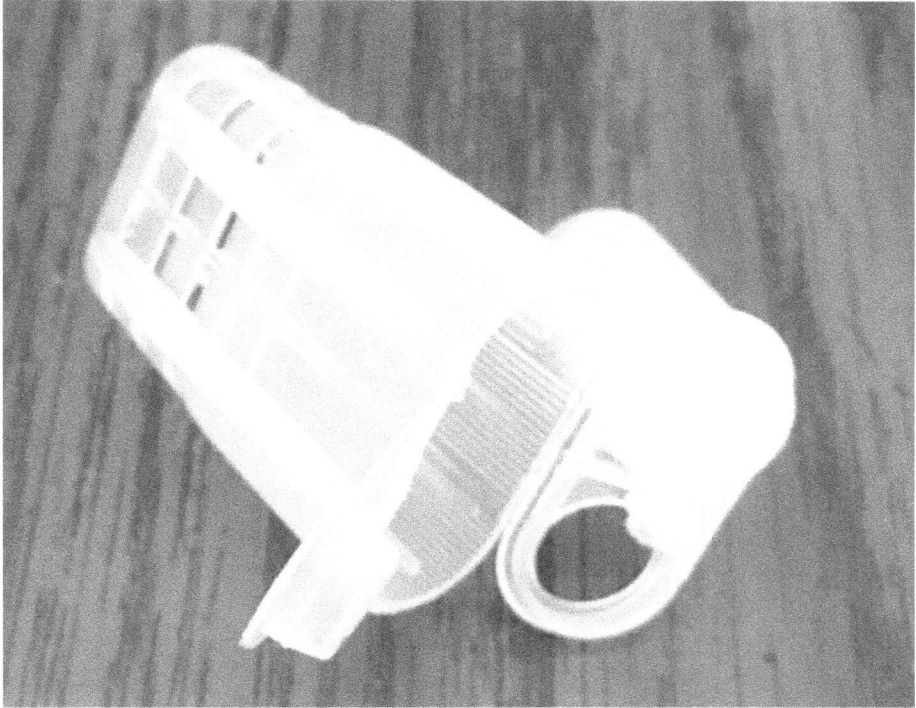

Candy

I guess someone somewhere sells queen candy. I have never bought it. I take a tablespoon or so of Karo white corn syrup and add dry granulated white sugar until I can't get any more to clump up with the syrup. Then I pop it in the microware until it bubbles a bit. Then I add a bit more sugar if it looks runny. Then I use a spoon and my fingers to force it into the tube on the JZBZ cage until it's backed and wipe off the excess.

Caging a queen

The best way to do this is to practice with drones and your bare hands. Catch the queen and put her in the cage. Immediately put your finger over the opening. Don't worry about getting stung, I've been doing this more than four decades and never had a queen sting me. Wait until she is facing away from the opening and quickly flip the tube and end closed and put your finger over the hole. Snap it. Now catch the attendants. Again, you could practice with drones, but you'll soon find you can catch a worker with its head in a cell, by the wings and put its head into the hole. Wait a second for it to sense the opening and the queen and then let it run in. If you are too afraid to catch workers with your fingers, use a pair of tweezers and catch one by the wing and put its head in the hole then let it run in. You can do this even with gloves on. Keep your finger over the hole at all times except when you're going to put another worker in. Keep an eye on the queen. It's helpful if she is facing away from the hole as she may make a run for it. I usually put in four attendants as the JZBZ cages are rather small. If the weather is likely to be cold instead of hot I put in seven. I never put in less than four or more than seven attendants. After the attendants are in, close the cap over the hole. This also locks the top closed.

Shipping

I don't really like any of my options for shipping so it's the best I can find at the present time. I live in a small town and whether I ship a queen from here or from the big city she goes out at the same speed so I prefer to mail them from my local post office. There is no UPS or FedEx available in town. I would have to drive to the city and the UPS people I've talked to don't seem to know that they ship queens. So I ship my queens USPS Express mail. This gives me a guaranteed delivery date (usually but not always overnight) and a shipping refund if they are late. At present, they are usually late so often it costs me nothing but the trouble of filing the paperwork for a refund. It is a pain, but it is worth filing it. If they are late and that causes the death of the queen, I file for damages. This you can do online. Get the customer to take a picture of the dead queen. File the picture with the online paperwork. Proof of value is a link to my web page where my prices are listed. Express mail comes with some insurance already and if you are shipping a lot of queens you can buy more.

Weather

I never ship if the predicted weather either at my house or the destination is going to be higher than the upper 80s F. Lower 80s F seems to work ok, but if it's going to be above that I wait. I try to ship on Monday's but at least no later than Wednesday. That way if the shipment runs late the queen is still likely to make it by Saturday and not get left somewhere until Monday. I also never ship when the predicted temperature here or there is going to fall below 40 F. If the queen's temperature falls below 40 she will likely die or be infertile. But even if the weather falls to 40 she is unlikely to because the package is usually either in the back of an enclosed truck or in a building or in the cab of a truck being delivered. So likely she won't fall below 50 F even if the outdoor temps are 40 F. Also I make adjustments to the ventilation of the envelope.

Ventilation

I use a hole punch to put holes all around the edges. In hot weather I add two in the middle even though one side will be covered by the label. The other side will let some air into the middle. In cold weather I leave out these middle holes. I also may do about half as many around the edges in cold weather.

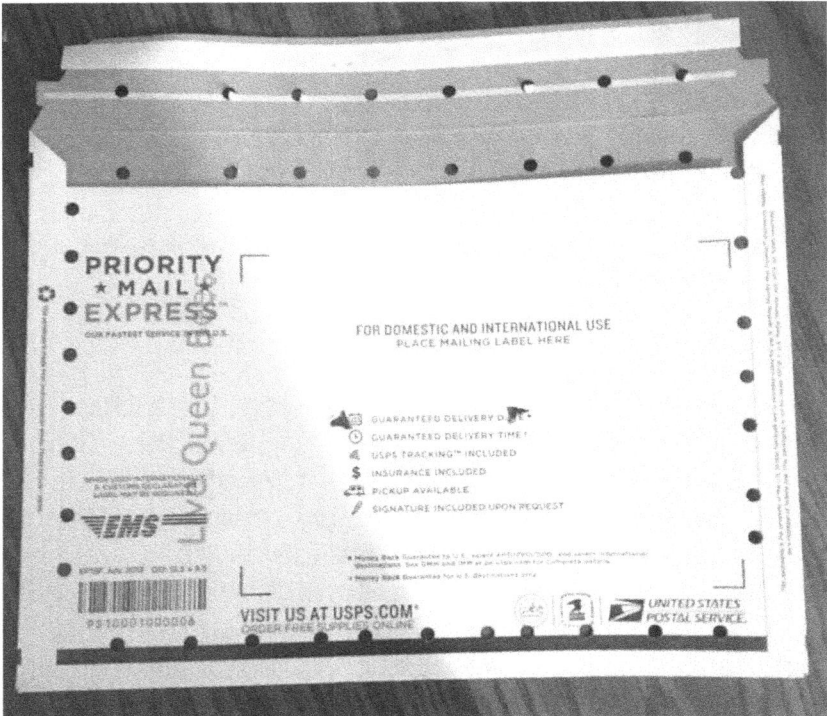

Mailing envelope

Communication

You should let the customer know you are shipping the queens. Give them the tracking number. Tell them it is best to inform their Post Office and give instructions as the what works best. Sometimes it works best to pick them up at the Post Office. Sometimes it works best to have them delivered to the house. Make sure you use the appropriate checkbox on the Express paperwork to require a signature if you want them to put it in the customer's hand. Keep in mind it may cost a day or so of time requiring a signature if that isn't going to happen quickly. Ask the customer what will work best.

Queen Banks

The concept of a queen bank is to keep a lot of mated queens until they are shipped or needed. I've had the best luck in setting up a queenless nuc, catching all the queens at the same time and putting them in the bank. Adding queens later disrupts things and usually a lot of queens are killed. So I use the queens until there are none. Meanwhile the bees usually raise a queen. Whenever I see that they have I remove that queen and put her in her own nuc somewhere. This often works out well because she has laid enough brood to keep the bank going. When I've removed all of the queens, I can catch another batch to put in the queen bank. I use JZBZ cages in a frame like this one that I got from Honey Run Apiaries:

One of my dilemmas has been whether or not to put attendants in with the queens and whether or not to have candy in the cage. The attendants, of course, won't live forever, but one of the issues is that a wax moth worm will get in the cage and the queen won't do anything about it. The wax moth will spin webs and the queen will be limited in their mobility and eventually get caught in the webs and die. With attendants this doesn't happen. The other problem is that the candy and the cage can get pretty funky if the queen is there for a while and there is candy in the cage. Of course the solution to both of these is not to keep them in the bank very long and that is usually my goal. I prefer to catch the queens on Saturday and Sunday and ship them on Monday. So I may as well have attendants in the cages.

The other issue, which of course is related, is how long can you bank them before it affects the quality of the queen. My conclusion after a lot of experimentation is that it doesn't matter how long you bank them but it does matter how *soon* you bank them. If you let them lay a couple of weeks before you bank them they will make fine queens no matter how long you bank them. But if you bank them too soon they will never make a good queen.

Winding Down in the Fall

When the bees don't make very many cells (or none) even though the starter started them, it's usually time to wind things down. As I catch queens rather than leave a mating nuc queenless I combine several queenless ones with a queenright one in my regular boxes. I try to end up with two eight frame medium boxes full of bees and stores as I consolidate. I try to get the brood together and the honey together but with mating nucs each frame is often a mixture of both. I could, if I wanted to, just give the queenless nuc frames and bees to other colonies, boosting any that are weak. As I empty the mating nucs I need to put away the equipment. If the consolidated nucs don't have enough stores I need to feed them to get them heavy enough or steal honey from a strong hive. I need to make sure they are strong enough for winter.

Swarm

Things I don't currently do

I want to clarify some things people often think they need to do that I have tried and abandoned.

Cloake board

I have used one (in a sense, since it did not incorporate an excluder in the board but used one separately a box below the board). I have shown it and briefly mentioned it in a previous chapter. It's not a bad system, I just don't find it as reliable as a swarm box/starter hive and I often have a use for those bees I shake into the starter to boost failing mating nucs. So for the amount of work and the reliable outcomes I don't use one. The object is to easily change a hive from a queenless starter to a queenright finisher and to route extra bees to the starter. You could accomplish the same thing with a cover and a bottom board but you would have to unstack a box or two to make it queenright again.

Priming the cups with royal jelly

When I started I had read Doolittle's *Scientific Queen Rearing* and Smith's *Queen Rearing Simplified* and both of them primed the cups with royal jelly so of course I tried it. It's not that it didn't work ok, but it didn't work any better than not priming. Jay Smith who explains how to do it in *Queen Rearing Simplified* in 1923, has changed his mind by the time of *Better Queens* in 1949. He says:

"We used to prime our cells with bee milk but, after careful examination, believe it was a detriment, for the first thing the bees do is to remove all the

milk we had put in. Grafting in bare cells is better-
or rather not so bad."—Jay Smith, Better Queens

Double grafting

The idea of Double grafting is to get a nice pool of royal
jelly after grafting and then remove those larvae and
replace them with new younger larva. The idea is that
this will result in them being better fed. In my experience if you put the brood you are going to graft from
into the cell starter for one or two hours they will be
swimming in royal jelly and if you graft those using the
Chinese grafting tool you will pick up that royal jelly
when you graft. Again Jay Smith used to double graft
but by the time he wrote better queens he said this:

"In order to get the cells filled with bee milk the
same as they are when built during swarming, I al-
lowed the larvae in the grafted cells to remain for
two days till there was plenty of bee milk in them,
then removed the larvae and put in young larvae. I
hoped to get fine cells in this manner but the bees
seemed to think otherwise. They accepted but a
few of the cells and in some cases the larva was
pushed over to one side of the cell and the bee
milk all removed. In a few cases the bees accepted
the cells but placed a little thin milk on top of the
milk already in the cells. The few queens reared
were no better than the ones reared by grafting in
the regular way. This leads me to believe the dried
bee milk is not suitable for larva food but rather is
the crumbs left over from the feast. I believe the
real food is the very thin milk that is fed to the lar-
va. Then one may ask what is the advantage of

having so much dried milk left in the cells. That merely indicates that the growing larva was fed in a lavish manner which is very necessary if quality of queens is desired. One queen reared as described above performed in a manner I never knew one to perform before or since. She laid drone eggs only but none of them died in the cells as is common with drone layers but all matured into perfect drones and that queen fairly filled that hive with beautiful drones. Scientists who have made a study of the subject tell us that bee milk is the same in all stages. I am inclined to doubt this. It may be changed in the moisture content only but I have observed that bee milk in the cell of a larva of one age will not be accepted by a larva of another age. When given, the bees immediately remove it and proceed to give the proper nourishment."—Jay Smith, Better Queens

---

Cell protectors

The idea of a cell protector is to keep the bees to whom you give the queen cell, from tearing down the cell. As far as I know this was invented by Doolittle. I found them to be useless. If the bees don't want the queen they will kill her when she emerges. If you want to keep them from rejecting a queen, make sure you aren't in a dearth. In a dearth they often tear down cells. In a flow they don't. If you aren't in a flow you may have to feed. I will defer to Jay Smith again:

"Why Bees Tear Down Cells

"Well-fed bees very seldom tear down cells. Just why I do not claim to know. I discovered this fact about thirty years ago. We were then using a mating hive holding two frames of honey. The bees tore down the cells just as fast as we gave them. I had been told that bees tear down cells because they are strange to them, similar to their reaction when introducing a strange queen. If that were true it seemed that little could be done about it. I sat down on a stump to think it over for most things can be worked out if we use the right formula. I remember that bees tear down cells worse at times than others. Why? It could not be because the cells were strange to the bees for they were always strange yet at times few were torn down while at others, as at present, nearly all were destroyed. Another reason was this: In introducing cells I often had some left over and would put them back into any hive I was using for cell building, often putting them into a hive that had not built them, yet in doing this many times, there never was a single cell torn down. That was proof that the bees did not tear them down because they were strange. Then why did the bees in the mating hives destroy the cells while the bees in the cell builders did not? Plainly it was because the bees in

the cell-building colonies have been well fed while the bees in the mating hives had not. To test out my theory I fed the bees in the mating hives that were to receive the cells the next day. The more I thought of it the more certain I was that it would work. But, what would the bees think about it? Cells were given the next day and I was delighted to see a high percentage of them was accepted. Well-Fed Bees do not Tear Down Cells

"So my conclusion was that well-fed bees do not tear down cells. True, even during a heavy honey flow some cells will be destroyed but in such cases there may be a large amount of unsealed brood and few fielders so in reality the bees are not well fed. Liberal feeding will make acceptance sure."—Jay Smith, Better Queens

Incubators

Maybe I shouldn't say I *don't* use them but I seldom use them. The results of using them has not been spectacular nor has introducing virgin queens. The concept is that the incubator will free up resources so you don't need as many resources to manage your queen cells. But I have much better luck in the hive with the bees doing the job. Also I have much better luck with queens that emerge in the mating nuc than in introducing virgin queens to the mating nucs.

Hair curler cages

Again, I can't say I *never* use them, but I seldom do as mentioned above I have never had really good luck introducing virgins. Maybe if I worked at tweaking the details of the process I could get where my success rate was good enough to do that, but so far I haven't been that successful with virgins. The idea of a "Hair curler" cage is to have the queen emerge into the cage, often in an incubator and then introduce the virgin queen to the mating nuc. If you do this, make sure the emerging queen has some access to some crystallized honey. The bigger ones have some recesses in the bottom and I fill those with crystallized honey. Honey, so she doesn't starve, and crystallized so she doesn't get all sticky from the honey. But then you're still stuck introducing virgin queens.

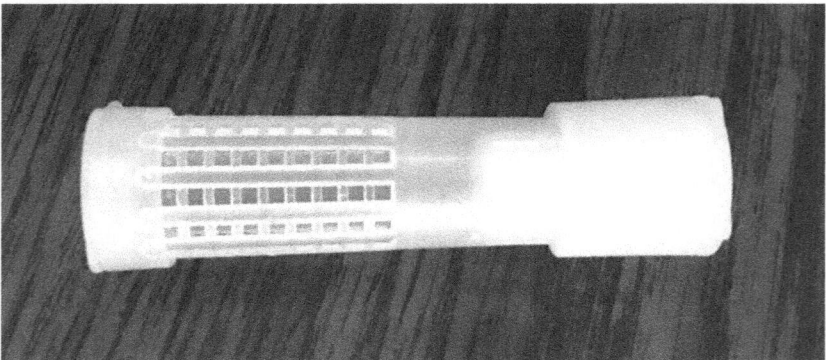

Accelerated queen rearing

This is another one of those "I can't say I *never...*" things. When I have some left over cells and not enough mating nucs or resources to stock mating nucs I may do this. The idea of this is to put a cell in the first week, but then the next week, before that queen is mated but after she has emerged, you put a cell in a hair curler cage or some variant of that so that another virgin emerges into the mating nuc every week. My first problem with this is that I really want a queen to lay for at least two weeks before I remove her and she often doesn't even *start* laying until two weeks after the cell is put in. But my other issue is that one of the two often ends up dead. In other words my success rate is not high enough to do it on a regular basis.

Working mating nucs

Synopsis

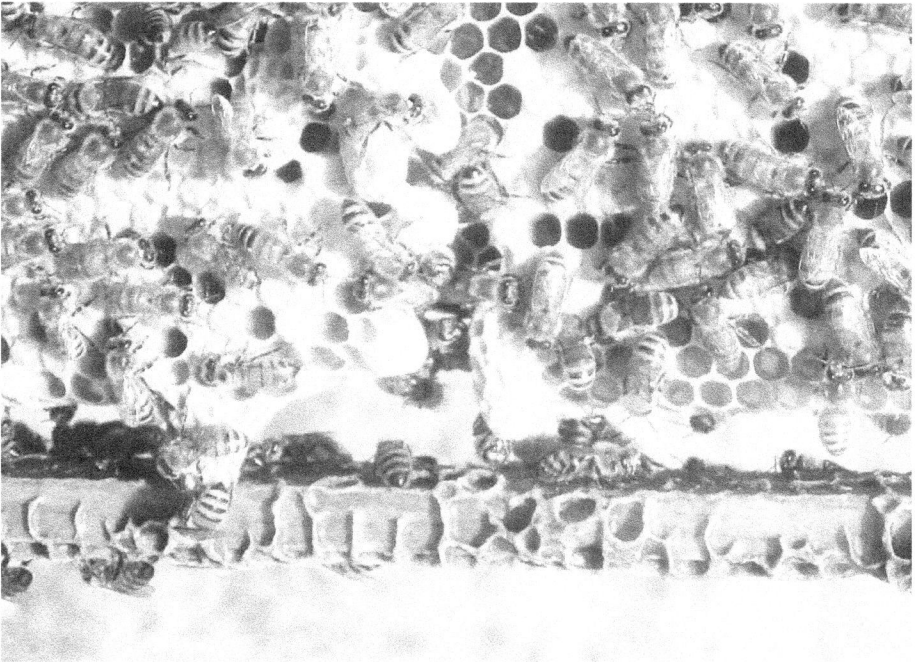

Queen cells

So a rough outline of what we have detailed is this:

• Soak some sponges and put them in the bottom of the starter.

• Put two frames of nectar/pollen in the starter box.

• Shake bees into starter until it is packed with bees

• Add two more frames of nectar/pollen for a total of two frames of bee bread (pollen) and two frames of nectar. If you can't find nectar, put in a two frames of honey and scratch the cappings a bit.

• Find some of the right age larvae in the breeder queen's hive and put them in the starter hive and put the starter hive in the shade.

• Come back in an hour or two and get the larvae frames and go back to the house to graft.

• Put the grafted cells in the starter hive and close it up again.

• Put the larvae frames back in the breeder hive or anywhere else you think you would rather have them.

• Set up a cell finisher hive.

• Start going through mating nucs from the last round and the round before that looking for and marking nucs that are queenless or the cell didn't emerge. Catching queens you are going to ship. Boosting nucs that are weak.

• Put cells from the last round of queens into the mating nucs from last week's cell starter.

• If you need more mating nucs set them up with a frame of brood, a frame of honey and a shake of bees.

• If the weather is hot put the cells from the starter into the finisher 24 hours later. If the weather is cool put the cells from the starter into the finisher 48 hours later.

These processes may take two or three days to complete. Next week we repeat them all

Conclusion

In conclusion I will defer to G.M. Doolittle who in 1845 wrote:

"At the outset, I shall undoubtedly be met by those inevitable "Yankee questions" - Does Queen-Rearing pay? Would it not pay me better to stick to honey-production, and buy the few queens which I need, as often as is required?

"I might answer, does it pay to kiss your wife? to look at anything beautiful? to like a golden Italian Queen? to eat apples or gooseberries? or anything else agreeable to our nature? is the gain in health, strength, and happiness, which this form of recreation secures, to be judged by the dollar-and-cent stand-point of the world?

"Can the pleasure which comes to one while looking at a beautiful Queen and her bees, which have been brought up to a high stand-point by their owner, be bought? Is the flavor of the honey that you have produced, or the keen enjoyment that you have had in producing it, to be had in the market?

"In nothing more than in Queen-Rearing, can we see the handiwork of Him who designed that we should be climbing up to the Celestial City, rather than groveling here with a "muck-rake" in our hands (as in "Pilgrim's Progress"), trying to rake in the pennies, to the neglect of that which is higher and more noble. There is something in working for

*better Queens which is elevating, and will lead one
out of self, if we will only study it along the many
lines of improvement which it suggests. I do not
believe that all of life should be spent in looking after the "almighty dollar;" nor do I think that our
first parents bustled out every morning, with the
expression seen on so many beekeepers' faces,
which seem to say, "Time is Money" The question,
it seems to me, in regard to our pursuit in life,
should not be altogether, "How much money is
there in it?" but, "Shall we enjoy a little bit of Paradise this side of Jordan. However, being aware, of
the general indifference to Paradise on either side
of Jordan, I will state that I have made Queen-
Rearing pay in dollars and cents..."*

Doing a cutout

Appendices

Rubber banding frames during a cutout

Glossary

Note: many of these terms are Latin and the plural of the ones with an "a" ending will be "ae". The plural of the "us" endings will be "i". Also meanings are given in the context of beekeeping. Also I have pared this down to things I think are directly or only somewhat indirectly related to queen rearing.

A

Africanized Honey Bees = I have heard these called Apis mellifera scutelata, but Scutelata are actually African bees from the Cape. They used to be called Adansonii, at least that's what Dr. Kerr, who bred them, thought they were. AHB are a mixture of African (Scutelata) and Italian bees. They were created in an attempt to increase production of bees. The USDA bred these at Baton Rouge from stock obtained from Dr. Kerr in Brazil. The USDA shipped these queens to the continental US over the course of many years. The Brazilians also were experimenting with them and the migration of those bees has been followed in the news for some time. They are extremely productive and extremely vigorous bees that are extremely defensive.

Alley Method = A graftless method of queen rearing system where bees are put in a "swarm box" to convince them of their queenlessness and a strip of old brood comb is cut and put on a bar for the bees to build into queen cells.

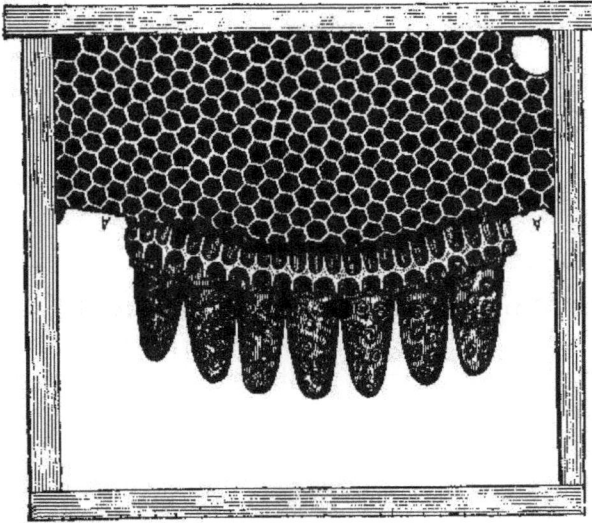

Alley Method

American Foulbrood = For more detail see the chapter on *Enemies of the Bees.* Caused by a spore forming bacteria. It used to be called Bacillus larvae but has recently been renamed Paenibacillus larvae. With American foulbrood the larvae usually dies after it is capped The brood pattern will be spotty. Cappings will be sunken and sometimes pierced. Recently dead larvae will string when poked with a matchstick. The smell is rotten and distinctive. Older dead larvae turn to a scale that the bees cannot remove.

Attendants = Worker bees that are attending the queen. When used in the context of queens in cages, the workers that are added to the cage to care for the queen.

Apis mellifera mellifera = These are the bees native to England or Germany. They have some of the characteristics of the other dark bees. They tend toward being runny (excitable on the combs) and a bit swarmy, but also seem to be well adapted to damp Northern climates.

Apis mellifera = Includes the honey bees originating in Africa and Europe.

B

Bacillus larvae = The outdated name for Paenibacillus Larvae, the bacteria that causes American foulbrood.

Bacillus thuringiensis = A naturally occurring bacteria that is sprayed on empty comb to kill wax moths. Also sold to control larvae of other specific insects.

Bait hive aka Decoy hive aka Swarm trap = A hive placed to attract stray swarms. Optimum bait hive: At least 20 liters of volume. 9 feet off the ground. Small entrance. Old comb. Lemongrass oil. Queen substance.

Balling = Worker bees surrounding a queen either to confine her because they reject her or to confine her to protect her.

Banking queens = Putting multiple caged queens in one nuc or hive.

Bearding = When bees congregate on the front of the hive.

Bee bread = Fermented pollen stored in the hive to use to feed brood.

Bee brush = Soft brush or whisk or large feather or handful of grass used to remove bees from combs.

Bee gum = A piece of a hollow tree used for a hive.

Bee jacket = A white jacket, usually with a zip on veil and elastic at the sleeves and waist, worn as protection when working bees.

Bee space = A space between $^1/_4$ and $^3/_8$ inch which permits free passage for a bee but too small to encourage comb building, and too large to induce propolizing.

Bee suit = A pair of white coveralls made for beekeepers to protect them from stings and keep their clothes clean. Most come with zip-on veils.

Bee tree = A hollow tree occupied by a colony of bees.

Bee vac aka Bee vacuum = A vacuum used to suck up bees when doing a cutout or removal. Usually converted from a shop vac. It needs careful adjustment to not kill the bees.

Bee venom = The poison secreted by special glands attached to the stinger of the bee which is injected into the victim of a sting.

Beehive = A box usually with movable frames, used for housing a colony of bees.

Beelining = Finding feral bees by establishing the line which the bees fly back to their home. This can also include marking and timing the bees to get the distance and triangulating the location by releasing the bees from various places.

Beeswax = A substance that is secreted by bees by special glands on the underside of the abdomen, deposited as thin scales, and used after mastication and mixture with the secretion of the salivary glands for constructing the honeycomb. The melting point of beeswax is 144 to 147 °F.

Better Queens method = A graftless queen rearing method similar to Isaac Hopkins' actual queen rearing method (as opposed to the "Hopkins Method"). Sort of the Alley Method but with new comb instead of old.

Betterbee = A beekeeping supply company out of New York. They have many things no one else does. They also have eight frame equipment.

Black scale = Refers to dried pupa, which died of American foulbrood.

Boardman feeder = A feeder that goes in the entrance and hold an inverted quart mason jar. They are notorious for causing robbing.

Bottom board = The floor of a bee hive.

Bottom board feeder = This is picture of the bottom board feeder that Jay Smith came up with. It's simply a dam made with a $^3/_4''$ by $^3/_4''$ block of wood put an inch or so back from where the front of the hive would be (18" or so forward of the very back). The box is slid forward enough to make a gap at the back. The syrup is poured in the back. A small board can be used to block the opening in the back. The bees can still get out the front by simply coming down forward of the dam. The picture is from the perspective of standing behind the hive looking toward the front. The edges of the dam have been enhanced and labels put on to try to make more sense. This version doesn't work on a weak hive as the syrup is too close to the entrance. It drowns as many bees as the frame feeders.

Breeder hive = The hive from which eggs or larvae are taken for queen rearing. In other words the donor hive.

Brood = Immature bees not yet emerged from their cells; in other words, egg, larvae or pupae.

Brood nest = The part of the hive interior in which brood is reared; usually the two bottom boxes.

Buckfast = A strain of bees developed by Brother Adam at Buckfast Abbey in England, bred for disease resistance, disinclination to swarm, hardiness, comb building and good temper.

C

Candy plug = A fondant type candy placed in one end of a queen cage to delay her release.

Capped brood = Immature bees whose cells have been sealed over with papery caps.

Carniolan bees = Apis mellifera carnica. These are darker brown to black. They fly in slightly cooler weather and in theory are better in northern climates.

They are reputed by some to be less productive than Italians, but I have not had that experience. The ones I have had were very productive and very frugal for the winter. They winter in small clusters and shut down brood rearing when there are dearths.

Castes = The three types of bees that comprise the adult population of a honey bee colony: workers, drones, and queen

Caucasian bees = Apis mellifera caucasica. They are silver gray to dark brown. They do propolis excessively. It is a sticky propolis rather than a hard propolis. They often coat everything with this sticky kind of proplolis, like fly paper. They build up a little slower in the spring than the Italians. They are reputed to be more gentle than the Italians. Less prone to robbing. In theory they are less productive than Italians. I think on the average they are about the same productivity as the Italians, but since they rob less you get less of the really booming hives that have robbed out all their neighbors.

Cell = The hexagonal compartment of a honey-comb.

Cell bar = A wooden strip on which queen cups are suspended for rearing queen bees.

Cell cup = Base of an artificial queen cell, made of beeswax or plastic and used for rearing queen bees or an empty beginning of a queen cell that the bees often build for no reason.

Cell finisher = A hive used to finish queen cells i.e. take them from capped to just before emergence. Sometimes queenright, sometimes queenless.

Cell starter = A hive used to start queen cells i.e. take them from just grafted to capped. Sometimes a "swarm box" or sometimes just a queenless hive.

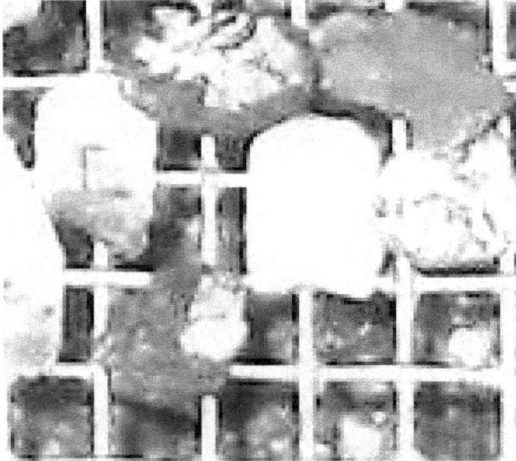

Chalkbrood = This is caused by a fungus Ascosphaera apis. It arrived in the US in 1968. If you find white pellets in front of the hive that kind of look like small corn kernels, you probably have chalkbrood. Putting the hive in full sun and adding more ventilation usually clears this up. Honey instead of syrup may contribute to clearing this up, since sugar syrup is much more alkaline (higher pH) than honey.

Chilled brood = Immature bees that have died from exposure to cold; commonly caused by mismanagement or sudden cold spells.

Chinese grafting tool = Grafting tool made of plastic, horn and bamboo that has a retractable "tongue" that slides under the larvae and, when released, pushes it off of the "tongue". Popular because it is easier to operate than most grafting needles and it lifts up more royal jelly in the process. Quality varies and most recommend buying several and picking the ones you like out of those.

Clipping = The practice of taking part of one or both wings off of a queen both for discouraging or slowing swarming and for identification of the queen.

Cloake Board AKA FWOF (Floor without a floor) = A device to divide a colony into a queenless cell starter and reunite it as a queenright cell finisher without having to open the hive.

Cloake board

Cluster = The thickest part of the bees on a warm day, usually the core of the brood nest. On a day below 50º F the only location where the bees are. It is used to refer both to the location and to the bees in that location.

Cocoon = A thin silk covering secreted by larval honey bees in their cells in preparation for pupation.

Colony = The superorganism made up of worker bees, drones, queen, and developing brood living to-gether as a family unit.

Comb = The wax structures in a colony in which eggs are laid, and honey and pollen are stored. Shaped like hexagons.

Comb foundation = A commercially made structure consisting of thin sheets of beeswax with the cell bases of a particular cell size embossed on both sides to induce the bees to build a that size of cell.

Cordovan bees = A subset of the Italians. In theory you could have a Cordovan in any breed, since it's technically just a color, but the ones for sale in North American that I've seen are all Italians. They are slightly more gentle, slightly more likely to rob and quite striking to look at. They have no black on them and look very yellow at first sight. Looking closely you see that where the Italians normally have black legs and head, they have a purplish brown legs and head.

Crimp-wired foundation = Comb foundation into which crimp wire is embedded vertically during foundation manufacture.

Cupralarva = A particular brand of graftless queen rearing system.

Cut-out = Removing a colony of bees from somewhere that they don't have movable comb by cutting out the combs and tying them into frames.

D

Dadant = A beekeeping supply company out of Illinois. Founded by C.P. Dadant who was a pioneer in the modern beekeeping era and invented, among other things, the Jumbo and the square Dadant box. ($19^7/_8$" by $19^7/_8$" by $11^5/_8$"), published and wrote for the American Bee Journal and translated *Huber's Observations on Bees* from French to English and published many books including but not limited to the later versions of *The Hive and the Honey Bee*.

Dearth = A period of time when there is no available forage for bees, due to weather conditions (rain, drought) or time of year.

Decoy hive aka Bait hive aka Swarm trap = A hive placed to attract stray swarms.

Depth = The vertical measurement of a box or frame.

Dequeen = To remove a queen from a colony. Usually done before requeening, or as a help for brood diseases or pests.

Diploid = Possessing pairs of genes, as workers and queens do, as opposed to haploid, possessing single genes as drones do.

Disease resistance = The ability of an organism to avoid a particular disease; primarily due to genetic immunity or avoidance behavior.

Dividing = Separating a colony to form two or more colonies. AKA a split

Division = Separating a colony to form two or more colonies.

Division board = A wooden or plastic piece like a frame but tight all the way around used to divide one box into more compartments for nucs.

Division board feeder or Frame feeder = A wooden or plastic compartment which is hung in a hive like a frame and contains sugar syrup to feed bees. The original designation (Division) was because it was *used* to make a division between two halves of a box to divide it into nucs, usually for queen rearing or making increase (splits). Most of them have a beespace around them now and cannot be used to make a division.

Domestic = Bees that live in a manmade hive. Since all bees are pretty much wild this is a relative term.

Doolittle method = A method of queen rearing that involves grafting young larvae into queen cups. First discovered by Nichel Jacob in 1568, then written about by Schirach in 1767 and then Huber in 1794 and finally popularized by G.M. Doolittle in his book *Scientific Queen Rearing* in 1846.

Double screen = A wooden frame, $^1/_2$ to $^3/_4$″ thick, with two layers of wire screen to separate two colonies within the same hive, one above the other. Often an entrance is cut on the upper side and placed to the rear of the hive for the upper colony and sometimes other openings are incorporated which would then be a Snelgrove board.

Double story or Double deeps = Referring to a beehive wintering in two deep boxes.

Drawn combs = Full depth comb ready for brood or nectar with the cell walls drawn out by the bees, completing the comb as opposed to foundation that has not been worked by the bees and has no cell walls yet.

Drifting = The movement of bees that have lost their location and enter hives other than their own home. This happens often when hives are placed in long straight rows where returning foragers from the center hives tend to drift to the row ends or when making splits and the field bees drift back to the original hive.

"The percentage of foragers originating from different colonies within the apiary ranged from 32 to 63 percent"—from a paper, published in 1991 by Walter Boylan-Pett and Roger Hoopingarner in Acta

Horticulturae 288, 6th Pollination Symposium (see Jan 2010 edition of Bee Culture, 36)

Drone = The male honey bee which comes from an unfertilized egg (and is therefore haploid) laid by a queen or less commonly, a laying worker.

Drone comb = Comb that is made up of cells larger than worker brood, usually in the range of 5.9 to 7.0mm in which drones are reared and honey and pollen are stored.

Drone brood = Brood, which matures into drones, reared in cells larger than worker brood. It is noticeably larger than worker brood and the cappings are distinctly dome shaped.

Drone Congregation Area = A place that drones from many surrounding hives congregate and wait for a queen to come. In other words a mating area. Drones find them by following both pheromone trails and topographical features of the landscape such as tree rows.

Drone layers = A drone laying queen (one with no sperm left to fertilize eggs) or laying workers.

Drone laying queen = A queen that can lay only unfertilized eggs, due to age, improper or late mating, disease or injury.

Drone mother hive = The hive which is encouraged to raise a lot of drones to improve the drone side of mating queens. Based on the myth that you can make bees raise more drones. Taking drone comb from the ones you want to perpetuate and giving them to other colonies is the only real way to succeed at this as the mother colony will then raise more drones while the colonies receiving the drone comb will raise less of their own because they will be raising the ones from the drone mother.

E

Eight frame = Boxes that were made to take eight frames. Usually between $13^1/_2$" and 14" wide depending on the manufacturer. Typically $13^3/_4$" wide.

Eggs = The first phase in the bee life cycle, usually laid by the queen, is the cylindrical egg $^1/_{16}$" (1.6 mm) long; it is enclosed with a flexible shell or chorion. It resembles a small grain of rice.

End bar = The piece of a frame that is on the ends of the frame i.e. the vertical pieces of the frame.

Entrance reducer = A wooden strip used to regulate the size of the entrance.

European Foulbrood = Caused by a bacteria. It used to be called Streptococcus pluton but has now been renamed Melissococcus pluton. European Foul Brood is a brood disease. With EFB the larvae turn brown and their trachea is even darker brown. Don't confuse this with larvae being fed dark honey. It's not just the food that is brown. Look for the trachea. When it's worse, the brood will be dead and maybe black and maybe sunk cappings, but usually the brood dies before they are capped. The cappings in the brood nest will be scattered, not solid, because they have been removing the dead larvae. To differentiate this from AFB use a stick and poke a diseased larvae and pull it out. The AFB will "string" two or three inches.

European Honey Bees = Bees from Europe as opposed to bees originating in Africa or other parts of the world or bees crossbred with those from Africa.

Ezi Queen = A particular brand of graftless queen rearing system.

F

Frame feeder or division board feeder = A wooden or plastic compartment which is hung in a hive like a frame and contains sugar syrup to feed bees. The original designation (Division) was because it was *used* to make a division between two halves of a box to divide it into nucs, usually for queen rearing or making increase (splits). Most of them have a beespace around them now and cannot be used to make a division.

Feeders = Any device used to feed bees.

Feral (queen or bees) = Since all North American bees are considered to have come from domestic stock, what most people call "wild" bees are really "feral" bees. Some use the term for survivor bees that were captured and used to raise queens meaning they *were* feral as opposed to *are* feral.

Fertile queen = An inseminated queen.

Fertilized = Usually refers to eggs laid by a queen bee, they are fertilized with sperm stored in the queen's spermatheca, in the process of being laid. These develop into workers or queens.

Festooning = The activity of young bees, engorged with honey, hanging on to each other usually to secrete beeswax but also in bearding and swarming..

Field bees = Worker bees which are usually 21 or more days old and work outside to collect nectar, pollen, water and propolis; also called foragers.

Floor Without a Floor AKA FWOF AKA Cloake Board = A device to divide a colony into a queenless cell starter and reunite it as a queenright cell finisher without having to open the hive.

Forage = Natural food source of bees (nectar and pollen) from wild and cultivated flowers. Or the act of gathering that food.

Foragers = Worker bees which are usually 21 or more days old and work outside to collect nectar, pollen, water and propolis; also called field bees.

Foundation = Thin sheets of beeswax embossed or stamped with the base of a worker (or rarely drone) cells on which bees will construct a complete comb (called drawn comb); also referred to as comb foundation, it comes wired or unwired and also in plastic as well as one piece foundations and frames as well as different thicknesses (thin surplus, surplus, medium) and different cell sizes (brood =5.4mm, small cell = 4.9mm, drone=6.6mm).

Foundationless = A frame with some kind of comb guide that is used without foundation.

Frame = A rectangular structure of wood designed to hold honeycomb, consisting of a top bar, two end bars, and a bottom bar; usually spaced a bee-space apart in the super.

Frame feeder = Sometimes called a "division board feeder". It takes the place of one or more frames. Less bees drown if you put floats in.

G

Gloves = Leather, cloth or rubber gloves worn while inspecting bees.

Grafting = Removing a worker larva from its cell and placing it in an artificial queen cup in order to have it reared into a queen.

Grafting tool = A needle or probe used for transferring larvae in grafting of queen cells.

H

Hair clip queen catcher = A device used to catch a queen that resembles a hair clip. Available from most beekeeping supply houses.

Haploid = Possessing a single set of genes, as drones do, as opposed to pairs of genes as workers and queens have.

Hive tool = A flat metal device used to pry boxes and frames apart, typically with a curved scraping surface or a lifting hook at one end and a flat blade at the other.

Hoffman frame = Frames that have the end bars wider than the top bars to provide the proper spacing when frames are placed in the hive. In other words, self-spacing frames. In other words, standard frames.

Hopkins method = A graftless method of queen rearing that involves putting a frame of young larvae horizontally above a brood nest.

Hopkins shim = A shim used to turn a frame flatways for queen rearing without grafting.

Hot (temperament) = Bees that are overly defensive or outright aggressive.

Hypopharyngeal gland = A gland located in the head of a worker bee that secretes "royal jelly". This rich blend of proteins and vitamins is fed to all bee larvae for the first three days of their lives and queens during their entire development.

I

Increase = To add to the number of colonies, usually by dividing those on hand. See Split.

Infertile = Incapable of producing a fertilized egg, as a laying worker or drone laying queen. Unfertilized eggs develop into drones.

Inner cover = An insulating cover fitting on top of the top super but underneath the outer cover, typically with an oblong hole in the center. Used to be called a "quilt board". In the old days these were often made of cloth.

Instar = Stages of larval development. A honey bee goes through five instars. The best queens are grafted in the 1st (preferably) or 2nd instar and not later than that.

Instrumental insemination aka II or AI = The introduction of drone spermatozoa into the spermatheca of a virgin queen by means of special instruments

Italian bees = A common race of bees, Apis mellifera ligustica, with brown and yellow bands, from Italy; usually gentle and productive, but tend to rob and brood incessantly.

J

Jenter = A particular brand of graftless queen rearing system.

L

Large Cell = Standard foundation size = 5.4mm cell size

Larva, open = The second developmental stage of a bee, starting the 4th day from when the egg is laid until it's capped on about the 9th or 10th day.

Larva, capped = The second developmental stage of a bee, ready to pupate or spin its cocoon (about the 10th day from the egg).

Laying workers = Worker bees which lay eggs in a colony caused by them being a few weeks without the pheromones from open brood; such eggs are infertile, since the workers cannot mate, and therefore become drones.

Leg baskets = Also called pollen baskets, a flattened depression surrounded by curved spines located on the outside of the tibiae of the bees' hind legs and adapted for carrying flower pollen and propolis.

Lemon Grass essential oil = Essential oil used for swarm lure which contains many of the constituents of Nasonov pheromone.

M

Marking = Painting a small dot of enamel on the back of the thorax of a queen to make her easier to identify and so you can tell her age and if she has been superseded.

Marking pen = An enamel pen used to mark queens. Available at local hardware stores as enamel pens. Also from beekeeping supply houses as Queen marking pens.

Marking Tube = A plastic tube commonly available from beekeeping supply houses that is used to safely confine a queen while you mark her.

Mating flight = The flight taken by a virgin queen while she mates in the air with several drones.

Mating nuc = A small nuc for the purpose of getting queens mated used in queen rearing.

Maxant = A beekeeping equipment manufacturer that makes uncappers, extractors, hive tools etc.

Medium = A box that is $6^5/_8''$ in depth and the frames are $6^1/_4''$ in depth. AKA Illinois AKA Western AKA $^3/_4$ depth.

Medium brood (foundation) = When used to refer to foundation, medium refers to the thickness of the wax *not* the depth of the frame. In this case it's medium thick and of worker sized cells.

Melissococcus pluton = New name given by taxonomists for the bacterium that causes European foulbrood. The old name was Streptococcus pluton.

Miller Bee Supply = A beekeeping supply company out of North Carolina *(www.millerbeesupply.com)*. Among other things, they have eight frame equipment.

Miller feeder = Top feeder popularized by C.C. Miller.

Miller Method = A graftless method of queen rearing that involves a ragged edge on some brood comb for the bees to build queen cells on.

N

Nasonov = A pheromone given off by a gland under the tip of the abdomen of workers that serves primarily as an orientation pheromone. It is essential to swarming behavior and nasonoving is set off by disturbance of the colony. It is a mixture of seven terpenoids, the majority of which is Geranial and Neral, which are a pair of isomers usually mixed and called citral. Lemongrass (Cymbopogon) essential oil is mostly these scents and is useful in bait hives and to get newly hived bees or swarms to stay in a hive.

Nasonoving = Bees who have their abdomens extended and are fanning the Nasonov pheromone. The smell is a mixture of lemon and geranium.

Natural cell = Cell size that bees have built on their own without foundation.

Natural comb = Comb that bees have built on their own without foundation.

Nectar = A liquid rich in sugars, manufactured by plants and secreted by nectary glands in or near flowers; the raw material for honey.

Nectar flow = A period of time when nectar is available.

New World Carniolans = A breeding program originated by Sue Cobey to find and breed bees from the US with Carniolan traits and other commercially useful traits.

Newspaper method = A technique to join together two colonies by providing a temporary newspaper barrier. Usually one sheet with a small slit. Usually you make sure both colonies can still fly and ventilate.

Nicot = A particular brand of graftless queen rearing system.

Nosema = Caused by a fungus (used to be classified as a protozoan) called Nosema apis. Nosema is present all the times and is really an opportunistic disease. The common chemical solution (which I don't use) was Fumidil which has been recently renamed Fumagilin-B. In my opinion the best prevention is to make sure your hive is healthy and not stressed and feed honey.

Nuc, nuclei, nucleus = A small colony of bees often used in queen rearing or the box in which the small colony of bees resides. The term refers to the fact that the essentials, bees, brood, food, a queen or the means to make one, are there for it to grow into a colony, but it is not a full sized colony.

Nurse bees = Young bees, usually three to ten days old, which feed and take care of developing brood.

O

Open Mesh Floor = British version of "screened bottom board".

Outyard = Also called out apiary, it is an apiary kept at some distance from the home or main apiary of a beekeeper.

Ovary = The egg producing part of a plant or animal.

Ovule = An immature female germ cell, which develops into a seed.

Ovariole = Any of several tubules that compose an insect ovary.

Oxytetracycline aka Oxytet = An antibiotic sold under the trade name Terramycin; used to control American and European foulbrood diseases.

P

Package bees = A quantity of adult bees (2 to 5 pounds), with or without a queen, contained in a screened shipping cage.

Parasitic Mites = Varroa and tracheal mites are the mites with economic issues for bees. There are several others that are not known to cause any problems.

Parthenogenesis = The development of young from unfertilized eggs laid by virgin females (queen or worker); in bees, such eggs develop into drones.

Phoretic = In the context of Varroa mites it refers to the state where they are on the adult bees instead of in the cell either developing or reproducing.

Piping = A series of sounds made by a queen, frequently before she emerges from her cell. When the queen is still in the cell it sounds sort of like a quack quack quack. When the queen has emerged it sounds more like zoot zoot zoot.

Play flights aka orientation flights = Short flights taken in front and in the vicinity of the hive by young bees to acquaint them with hive location; sometimes mistaken for robbing or swarming preparations.

Pollen = The dust-like male reproductive cells (gametophytes) of flowers, formed in the anthers, and important as a protein source for bees; fermented pollen (bee bread) is essential for bees to rear brood.

Pollen basket = An anatomical structure on the bees legs where pollen and propolis is carried.

Pollen bound = A condition where the brood nest of a hive is being filled with pollen so that there is nowhere for the queen to lay.

Pollen pellets or cakes = The pollen packed in the pollen baskets of bees and transported back to the colony made by rolling in the pollen, brushing it off and mixing it with nectar and packing it into the pollen baskets.

Pollen substitute = A food material which is used to substitute wholly for pollen in the bees' diet; usually contains all or part of soy flour, brewers' yeast, wheast, powdered sugar, or other ingredients. Research has shown that bees raised on substitute are shorter lived than bees raised on real pollen.

Pollen supplement = A mixture of pollen and pollen substitutes used to stimulate brood rearing in periods of pollen shortage

Pollen trap = A device for collecting the pollen pellets from the hind legs of worker bees; usually forces the bees to squeeze through a screen mesh, usually #5 hardware cloth, which scrapes off the pellets which fall through #7 hardware cloth into a drawer with a screened bottom so the pollen won't mold.

Prime swarm = The first swarm to leave the parent colony, usually with the old queen.

Proboscis = The mouthparts of the bee that form the sucking tube or tongue

Propolis = Plant resins collected, mixed with enzymes from bee saliva and used to fill in small spaces inside the hive and to coat and sterilize everything in the hive. It has antimicrobial properties. It is typically made from the waxy substance from the buds of the poplar family.

Propolize = To fill with propolis, or bee glue.

Pupa = The third stage in the development of the bee during which it is inactive and sealed in its cocoon.

Push In Cage = Cage made of #8 hardware cloth used to introduce or confine queens to a small section of comb. Usually used over some emerging brood.

Q

Queen = A fully developed female bee responsible for all the egg laying of a colony.

Queen Bank = Putting multiple caged queens in a nuc or hive.

Queen cage = A special cage in which queens are shipped and/or introduced to a colony, usually with 4 to 7 young workers called attendants, and usually a candy plug.

Queen cage candy = Candy made by kneading powdered sugar with invert sugar syrup until it forms a stiff dough; used as food in queen cages.

Queen cell = A special elongated cell resembling a peanut shell in which the queen is reared; usually over an inch in length, it hangs vertically from the comb.

Queen clipping = Removing a portion of one or both wings of a queen to prevent her from flying or to better identify when she has been replaced.

Queen cup = A cup-shaped cell hanging vertically from the comb, but containing no egg; also made artificially of wax or plastic to raise queens

Queen excluder = A device made of wire, wood or zinc (or any combination thereof) having openings of

.163 to .164 inch, which permits workers to pass but excludes queens and drones; used to confine the queen to a specific part of the hive, usually the brood nest.

Queenright = A colony that contains a queen capable of laying fertile eggs and making appropriate pheromones that satisfy the workers of the hive that all is well.

Queen Mandibular Pheromone aka Queen substance aka QMP = A pheromone produced by the queen and fed to her attendants who share it with the rest of the colony that gives the colony the sense of being queenright. Chemically QMP is very diverse with at least 17 major components and other minor ones. 5 of these compounds are: 9-ox-2-decenoic acid (9ODA) + cis & trans 9 hydroxydec-2-enoic acid (9HDA) + methyl-p-hydroxybenzoate (HOB) and 4-hydroxy-3-methoxyphenylethanol (HVA). Newly emerged queens produce very little of this. QMP is responsible for inhibition of rearing replacement queens, attraction of drones for mating, stabilizing and organizing a swarm around the queen, attracting a retine of attendants, stimulating foraging and brood rearing, and the general moral of the colony. Lack of it also seems to attract robber bees. You can save this by retiring your queens in a jar of alcohol instead of pinching them. You can also buy an artificial version sold as PseudoQueen by some of the beekeeping supply houses. These can be used as swarm lure or to settle a swarm or a cutout into their new home.

Queen muff = A screen wire tube that resembles a "muff" to keep your hands warm in shape but is used to keep queens from escaping when marking them or releasing attendants.

R

Rabbet = In wood working a groove cut into wood. The frame rests in a Langstroth hive are rabbets and the corners are sometimes done as rabbets and sometimes as finger or box joints.

Races of Bees = In taxonomy this is actually a variety but in beekeeping it is typically called a "race". All of these are Apis mellifera. The most common currently In the US are Italians (ligustica), Carniolans (carnica) and Caucasians (caucasica). Russians would be either carpatica, acervorum, carnica or caucasica depending on who you are talking to.

Regression = As applied to cell size, large bees, from large cells, cannot build natural sized cells. They build something in between. Most will build 5.1 mm worker brood cells. Regression is getting large bees back to smaller bees so they can and will build smaller cells.

Reorientation = When the bees take note of their surroundings and landmarks to make sure they remember the location of the colony. A variety of things set this off. Young bees will orient (not reorient but it's the same behavior) when they first emerge from the hive.

Requeen = To replace an existing queen by removing her and introducing a new queen.

Rendering wax = The process of melting combs and cappings and removing refuse from the wax.

Retinue = Worker bees that are attending the queen.

Robber screen = A screen used to foil robbers but let the local residents into the hive.

Robbing = The act of bees stealing honey/nectar from the other colonies; also applied to bees cleaning out wet supers or cappings left uncovered by beekeepers and sometimes used to describe the beekeeper removing honey from the hive.

Ropy = A quality of forming an elastic rope when drawn out with a stick. Used on capped brood as a diagnostic test for American foulbrood.

Rolling = A term to describe what happens when a frame is too tight or pulled out too quickly and bees get pushed against the comb next to it and "rolled". This makes bees very angry and is sometimes the cause of a queen being killed.

Royal jelly = A highly nutritious, milky white secretion of the hypopharyngeal gland of nurse bees; used to feed the queen and young larvae.

Russian bees = Apis mellifera acervorum or carpatica or caucasica or carnica. They came from the Primorsky region of Russia and were used for breeding mite resistance because they were already surviving the mites. They were brought to the USA by the USDA in June of 1997, studied on an island in Louisiana and then field testing in other states in 1999. They went on sale to the general public in 2000.

S

Scout bees = Worker bees searching for a new source of pollen, nectar, propolis, water, or a new home for a swarm of bees.

Self-spacing frames aka Hoffman frames = Frames constructed so that everything but the end bar (which is the spacer) is a bee space apart when pushed together in a hive body.

Shaken swarm = An artificial swarm made by shaking bees off of combs into a screened box and then putting a caged queen in until they accept her. One method for making a divide. Also the method used to make packages of bees.

Small Cell = 4.9mm cell size. Used by some bee-keepers to control Varroa mites.

Small Hive Beetle = A pest recently imported to North America, whose larvae will destroy comb and ferment honey.

Smith method = A method of queen rearing popularized by Jay Smith, that uses a swarm box as a cell starter and grafting larvae into queen cups.

Solar wax melter = A glass-covered box used to melt wax from combs and cappings using the heat of the sun.

Sperm cells = The male reproductive cells (gametes) which fertilize eggs; also called spermatozoa.

Spermatheca = A small sac connected with the oviduct of the queen bee in, which is stored, the spermatozoa received by the queen when mating with drones.

Spiracles = Openings into the respiratory system on a bee that can be closed at will. These are on the sides of the bee. They are considerably smaller than the Trachea they protect. The first thoracic spiracle is the one that is infiltrated by the tracheal mites as it is the largest. When closed the spiracles are air tight.

Split = To divide a colony for the purpose of increasing the number of hives.

Starter hive aka a Swarm box = A box of shaken bees used to start queen cells.

Streptococcus pluton = Deprecated (old) name for the bacterium that causes European foulbrood. The new name is Melissococcus pluton.

Sugar syrup = Feed for bees, containing sucrose or table (cane or beet) sugar and hot water in various ratios; usually 1:1 in the spring and 2:1 in the fall.

Sugar roll test = A test for Varroa mites that involves rolling a cupful of bees in powdered sugar and counting the number of mites dislodged.

Super = A box with frames in which bees store honey; usually placed above the brood nest. From the Latin *super* meaning "above".

Supering = The act of placing honey supers on a colony in expectation of a honey flow.

Supersedure = Rearing a new queen to replace the mother queen in the same hive.

Suppressed Mite Reproduction aka SMR = Queens from a breeding program by Dr. John Harbo that have less Varroa problems probably due to increased hygienic behavior. Lately renamed VSH aka Varroa Sensitive Hygiene.

Survivor stock = Bees raised from bees that were surviving without treatments. Often feral stock.

Swarm = A temporary collection of bees, containing at least one queen that split apart from the mother colony to establish a new one.

Swarm box aka a Starter hive = A box of shaken bees used to start queen cells.

Swarm cell = Queen cells usually found on the bottom of the combs before swarming.

Swarm commitment = The point just after swarm cutoff where the colony is committed to swarming.

Swarm cutoff = The point at which the colony decides to swarm or not.

Swarm trap aka Bait hive aka Decoy hive = A hive placed to attract stray swarms.

Swarm preparation = The sequence of activities of the bees that is leading up to swarming. Visually you can see this start at backfilling the brood nest so that the queen has nowhere to lay.

Swarming = The natural method of propagation of the honey bee colony.

Swarming season = The time of year, usually late spring to early summer, when swarms usually issue.

T

Tested queen = A queen whose progeny shows she has mated with a drone of her own race and has other qualities which would make her a good colony mother. One that has been given time to prove what her qualities are.

Thelytoky = A type of parthenogenetic reproduction where unfertilized eggs develop into females. Usually with bees this is referring to a colony rearing a queen from a laying worker egg. This is very rare, but documented, with European honey bees. It is common with Cape Bees.

Thorax = The central region of an insect to which the wings and legs are attached.

Tiger striped (queen) = Markings of a particular type on a queen. Not striped like a worker (who have very even bands) but more like "flames".

Top feeder = Miller feeder. A box that goes on top of the hive that contains the syrup. See Miller Feeder.

Tracheal Mites = A mite that infests the trachea of the honey bee. Resistance to tracheal mites is easily bred for.

Transferring or cut out = The process of changing bees and combs from trees, houses or bee gums or skeps to movable frame hives.

Trophallaxis = The transfer of food or pheromones among members of the colony through mouth-to-mouth feeding. It is used to keep a cluster of bees alive as the edges of the cluster collect food and share it through the cluster. It is also used for communication as pheromones are shared. One very important one is QMP (Queen Mandibular Pheromone) which is shared by trophallaxis throughout the hive.

U

Unfertilized = An ovum or egg, which has not been united with the sperm.

Uniting = Combining two or more colonies to form a larger colony. Usually done with a sheet of newspaper between.

Unlimited Brood Nest aka "food chamber" = running bees in a configuration where the brood nest is not limited by an excluder and they are usually over-wintered in more boxes to allow more food and more expansion in the spring.

V

Varroa destructor used to be called Varroa Jacobsoni = Parasitic mite of the honey bee.

Veil = A protective netting or screen that covers the face and neck; allows ventilation, easy movement and good vision while protecting the primary targets of guard bees.

Virgin queen = An unmated queen bee.

W

Warré hive = A type of vertical top bar hive invented by Abbé Émile Warré.

Washboarding = When the bees on the landing board or the front of a hive are moving in unison resembling a line dance.

Warming cabinet = An insulated box or room heated to liquefy honey or to heat honey to speed extraction.

Wax Dipping Hives = A method of protecting wood and also of sterilizing from AFB where the equipment is "fried" in a mixture of wax and gum resin. Usually done with paraffin sometimes done with beeswax.

Wax glands = The eight glands located on the last 4 visible, ventral abdominal segments of young worker bees; they secrete beeswax flakes.

Wax moths = See chapter *Enemies of the Bees*. Wax moths are opportunists. They take advantage of a weak hive and live on pollen, honey and burrow through the wax.

Wax scale or flake = A drop of liquid beeswax that hardens into a scale upon contact with air; in this form it is shaped into comb.

Wax tube fastener = A metal tube for applying a fine stream of melted wax to secure a sheet of foundation into a groove on a frame.

Western = I have seen this used in two ways. A box that is $6^5/_8$" in depth and the frames are $6^1/_4$" in depth. AKA Illinois AKA Medium AKA $^3/_4$ depth. Or referring to one that is $7^5/_8$".

Western Bee Supply = A beekeeping supply company out of Montana. The company that makes all of Dadant's equipment. Also sell eight frame equipment if you request it.

Windbreaks = Specially constructed, or naturally occurring barriers to reduce the force of the (winter) winds on a beehive.

Winter cluster = A tight ball of bees within the hive to generate heat; forms when outside temperature falls below 50º F.

Winter hardiness = The ability of some strains of honey bees to survive long winters by frugal use of stored honey.

Wire, frame = Thin 28# wire used to reinforce foundation destined for the broodnest or honey extractor.

Wire cone escape = A one-way cone formed by window screen mesh used to direct bees from a house or tree into a temporary hive.

Worker bees = Infertile female bee whose reproductive organs are only partially developed, and is anatomically different than a queen and is equipped and responsible for carrying out all the routine duties of the colony.

Worker comb = Comb measuring between 4.4mm and 5.4mm, in which workers are reared and honey and pollen are stored.

Worker Queen aka laying workers = Worker bees which lay eggs in a colony hopelessly queenless; such eggs are not fertilized, since the workers cannot mate, and therefore become drones.

Worker policing = Workers that remove eggs laid by workers.

Y

Yellow (queen or bees) = When used to refer to honey bees this refers to a lighter brown color. Honey bees are *not* yellow. A Yellow queen is usually a solid light golden brown.

Smoker

Acronyms

ABJ = American Bee Journal. One of the two main bee magazines in the USA.

AFB = American foulbrood

AHB = Africanized Honey Bees

AM = Apis mellifera. (European honey bees)

AMM = Apis mellifera mellifera

APV = Acute Paralysis Virus. This virus kills both adult bees and brood.

BC = Bee Culture aka Gleanings in Bee Culture. One of the two main Beekeeping magazines in the USA

BLUF = Bottom Line Up Front. A style of writing where you present the conclusion at the beginning. Common in scientific studies or military correspondence.

BPMS = Bee Parasitic Mite Syndrome

Carni = Carniolan = Apis mellifera carnica

Cauc = Caucasian = Apis mellifera Caucasia

CB = Checkerboarding (aka Nectar Management)

CCD = Colony Collapse Disorder

CPV = Chronic Paralysis Virus

CW = Conventional Wisdom

DCA = Drone Congregation Area

DVAV = Dorsal-Ventral Abdominal Vibrations dance.

DWV = Deformed Wing Virus

EAS = Eastern Apiculture Society

EFB = European Foulbrood

EHB = European Honey Bees

FGMO = Food Grade Mineral Oil

FWIW = For What It's Worth.

FWOF = Floor With Out a Floor

HAS = Heartland Apiculture Society

HBH = Honey Bee Healthy

HBTM = Honey Bee Tracheal Mite

HFCS = High Fructose Corn Syrup. A common bee feed.

HSC = Honey Super Cell (Fully drawn plastic comb in deep depth and 4.9mm cell size)

HMF = Hydroxymethyl furfural. A naturally occurring compound in honey that rises over time and rises when honey is heated.

IAPV = Israeli Acute Paralysis Virus. The virus currently being blamed for CCD

IPM = Integrated Pest Management

IMHO = In My Humble Opinion

IMO = In My Opinion

IMPOV = In My Point Of View

KTBH = Kenya Top Bar Hive (one with sloped sides)

KBV = Kashmir Bee Virus

LC = Large Cell (5.4mm cell size)

LGO = Lemon Grass (essential) Oil (used for swarm lure)

MAAREC = Mid-Atlantic Apiculture Research and Extension Consortium

NM = Nectar Management (aka Checkerboarding)

NWC = New World Carniolans

OA = Oxalic Acid. An organic acid used to kill Varroa as either a syrup or vaporized.

OSR = Oil Seed Rape (aka Canola). A crop that produces honey that is grown to produce oil.

PC = PermaComb (Fully drawn plastic comb in medium depth and about 5.0mm cell size)

PDB = Para Dichloro Benzene (aka Paramoth wax moth treatment)

PMS = parasitic mite syndrome

QMP = Queen Mandibular Pheromone

SBB = Screened Bottom Board

SBV = Sac Brood Virus

SC = Small Cell (4.9mm cell size)

SHB = Small Hive Beetle

SMR = Suppressed Mite Reproduction (usually referring to a queen)

TBH = Top Bar Hive

TM = Terramycin or Tracheal Mites depending on the context

T-Mites = Tracheal Mites

TTBH = Tanzanian Top Bar Hive (one with vertical sides)

ULBN = Unlimited Brood Nest

VD = Varroa destructor

VJ = Varroa jacobsoni

V-Mites = Varroa Mites

VSH = Varroa Sensitive Hygiene. Similar to and appears to be a more specific name for the SMR trait. A trait in queens that is being bred for where the workers sense Varroa infested cells and clean them out.

Index

About the Author

"His writing is like his talks, with more content, detail, and depth than one would think possible with such few words...his website and PowerPoint presentations are the gold standard for diverse and common sense beekeeping practices."—Dean Stiglitz

Michael Bush is an internationally recognized author and speaker on natural beekeeping. He is the author of The Practical Beekeeper, which has been published in five languages and is one of the authoritative sources on sustainable beekeeping without chemical treatments. His website is an invaluable source of information on natural, chemical-free beekeeping and just plain beekeeping. The topics are based on the questions that have come up on beekeeping forums over and over. The wisdom of his methods is evidenced by the fact that he successfully keeps bees through the harsh winters of Nebraska.

He has had an eclectic set of careers from printing and graphic arts, to construction to computer programming and a few more in between. Currently he is working in computers. He has been keeping bees since the mid 1970's, usually from two to seven hives up until the year 2000. Varroa forced more experimentation which required more hives and the number has grown steadily over the years from then. By 2008 it was about 200 hives. He is active on many of the Beekeeping forums with last count at about 60,000 posts between all of them. He has an extensive web site on beekeeping at www.bushfarms.com/bees.htm

---

"What Michael reminds us of, is that beekeeping is a dynamic art. And that like most things of beauty and meaning, they are part of a system, a dance of relationships that requires as much or more intuitive insight as hard science to understand." --Scott Klein (St. Louis Beekeepers)
